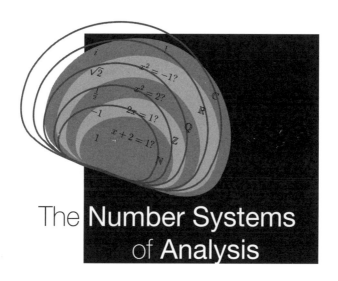

The Number Systems of Analysis

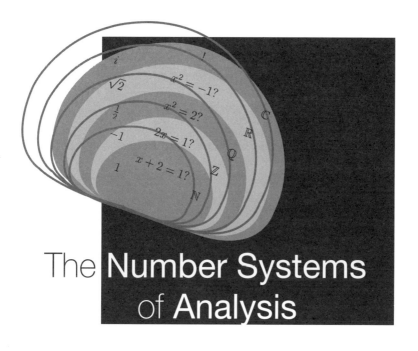

The Number Systems of Analysis

C H C Little • K L Teo • B van Brunt

Massey University, New Zealand

World Scientific
New Jersey • London • Singapore • Hong Kong

Published by

World Scientific Publishing Co. Pte. Ltd.
5 Toh Tuck Link, Singapore 596224
USA office: Suite 202, 1060 Main Street, River Edge, NJ 07661
UK office: 57 Shelton Street, Covent Garden, London WC2H 9HE

British Library Cataloguing-in-Publication Data
A catalogue record for this book is available from the British Library.

THE NUMBER SYSTEMS OF ANALYSIS

Copyright © 2003 by World Scientific Publishing Co. Pte. Ltd.

All rights reserved. This book, or parts thereof, may not be reproduced in any form or by any means, electronic or mechanical, including photocopying, recording or any information storage and retrieval system now known or to be invented, without written permission from the Publisher.

For photocopying of material in this volume, please pay a copying fee through the Copyright Clearance Center, Inc., 222 Rosewood Drive, Danvers, MA 01923, USA. In this case permission to photocopy is not required from the publisher.

ISBN 981-238-606-8

Printed in Singapore.

Preface

This book lays a firm foundation for the development of analysis. No specific mathematical knowledge on the part of the reader is assumed — the subject is developed from scratch — but a modicum of mathematical maturity is expected. However, the guiding principle of assuming nothing has forced us to treat carefully the introduction of several concepts already familiar to the reader in an intuitive sense, and it has seemed to us consistent with this approach to present arguments in somewhat more detail than is commonly the case in similar books. The authors and the reader must alike be satisfied that there is no lacuna in the logic. For this reason also no result upon which the development depends has been banished to the exercises, and proofs have been omitted only when the result is immediate or a reference to a virtually identical proof has been given in the book.

The most familiar number systems, in order from the simplest to the most complicated, are those of the natural numbers, integers, rational numbers, real numbers and complex numbers. The natural numbers are those used to count, and each of the other kinds of numbers are introduced because of the need to solve polynomial equations. For this purpose the complex numbers suffice, and the development of a yet more complicated number system is therefore unnecessary. This book traces the development of the number systems and demonstrates the inadequacy of each save the complex. The establishment of the adequacy of the complex numbers for solving polynomial equations provides a suitable goal for the book. The resulting theorem, called the fundamental theorem of algebra, is often left without a proof in works on the foundations of analysis. In the spirit of this book, however, it is inappropriate merely to state this theorem. We thus prove it but without introducing much algebraic or analytical machinery that is useful only to obtain the proof.

The reader might understandably query why it is necessary to study these number systems in such detail; the pragmatic lecturer might ponder whether any pedagogical advantage for the undergraduate student is gained by such a study. Aside from the intrinsic value of learning this foundational subject, it is a topic accessible to undergraduates that provides a substantial close-knit interplay of logic, set theory, algebra and analysis. Moreover, the construction of the real numbers provides an introduction to key concepts in analysis that generalize to more abstract settings.

The first half of the book has a decidedly algebraic flavour. The elements of logic that underpin the arguments in the book are introduced in Chapter 1. For our purposes logic is deemed to be a tool rather than a subject of inquiry in its own right. The treatment is therefore informal, but the need to avoid paradoxes has led the authors to be careful to explain which statements are to be deemed propositions and how propositions may be combined. Similar comments apply to the development of set theory in Chapter 2. The natural number system is constructed in Chapter 3. The integers and rational numbers are then developed efficiently using algebraic structures in the next two chapters. The construction of the real number system, however, requires fundamentally different concepts to augment the algebra, and it is here that the analysis begins in earnest with the introduction of sequences of rational numbers and convergence in Chapter 6.

The real number system is developed using Cauchy sequences in Chapter 7. Although different constructions are available (e.g. Dedekind cuts) we eschew these approaches and focus on the real numbers as the completion of the rational numbers in the sense of Cauchy sequence convergence. The student normally first encounters such a completion process in the framework of integration theory, where a space of Riemann integrable functions is "completed" by a space of Lebesgue integrable functions. This process mimics that used to construct the real numbers, but a considerable amount of analytical machinery is needed, and a fundamentally different (and complicated) definition of the integral is required. The real numbers thus provide a simpler introduction to the concept of completion and serve as a model for the completion of more abstract spaces.

The development of the number systems finishes with the appearance of complex numbers in Chapter 8. Our account culminates in Chapter 9 with a proof of the fundamental theorem of algebra using a blend of algebra and analysis.

The authors appreciate the patience and encouragement of their wives. They thank Fiona Davies for her help in preparing part of the manuscript and drafting the figures.

Charles Little
Kee Teo
Bruce van Brunt
Massey University, Palmerston North, New Zealand

Contents

Preface   v

1. **Elements of Logic**   1

    1.1 Introduction 1
    1.2 Propositions 3
    1.3 Predicates and Quantifiers 16

2. **Sets**   21

    2.1 Construction of Sets 21
    2.2 Ordered Pairs, Relations and Functions 31

3. **Natural Numbers**   43

    3.1 Definition and Basic Properties 43
    3.2 Addition 51
    3.3 Multiplication 53
    3.4 Exponentiation 56
    3.5 Order 57
    3.6 Applications 61

4. **Integers**   73

    4.1 Definition 73
    4.2 Addition 74
    4.3 Multiplication 76
    4.4 Rings 79
    4.5 Exponentiation 84
    4.6 Order 85

	4.7 Square Roots and Absolute Values	89
	4.8 An Extension of Induction	91
	4.9 Sums and Products	92
5.	**Rational Numbers**	**107**
	5.1 Definition	107
	5.2 Definition of Addition and Multiplication	108
	5.3 Fields	112
	5.4 Exponentiation	113
	5.5 Order	116
	5.6 The Archimedean Property and Certain Technical Results	119
6.	**Sequences on $\mathbb{Q}$**	**125**
	6.1 Sequences and Limits	125
	6.2 Cauchy Sequences	132
	6.3 Operations with Cauchy Sequences	133
	6.4 A Divergent Cauchy Sequence	137
7.	**The Real Number System**	**141**
	7.1 The Real Numbers	141
	7.2 Sequences on $\mathbb{R}$	146
	7.3 General Results Concerning Sequences	152
	7.4 Sets in $\mathbb{R}$	156
	7.5 Bounded Sets in $\mathbb{R}$	160
	7.6 Monotonic Sequences	167
	7.7 The Bolzano-Weierstrass Theorem	174
	7.8 The Contraction Mapping Principle	178
8.	**Complex Numbers**	**187**
	8.1 Definition	187
	8.2 Addition and Multiplication	187
	8.3 Conjugates	191
	8.4 Modulus	192
	8.5 Complex Sequences	194
	8.6 Sets in $\mathbb{C}$	199
	8.7 Quadratic Equations	202
9.	**The Fundamental Theorem of Algebra**	**207**

9.1 General Results Concerning Polynomials 207
9.2 Polynomials with Real Coefficients 212
9.3 nth Roots of a Complex Number 215
9.4 The Fundamental Theorem of Algebra 217

Index   221

Chapter 1

Elements of Logic

1.1 Introduction

Mathematics deals primarily with statements which can in principle be determined to be true or false. In this chapter we shall call such statements **propositions**. A proposition is therefore necessarily an unambiguous objective statement, devoid of any subjective or emotional content. For example, "a rose is a flower" and "a rose is an animal" are both propositions (the former being true and the latter false), but "a rose is beautiful" is not a proposition because of its lack of objectivity.

Sometimes the decision as to whether a given statement is a proposition requires more subtlety than was displayed in the preceding examples. For instance, consider the following statement: "this statement is false". If that statement were a proposition, it would be either true or false. However the possibility of it being true is ruled out by the fact that it asserts its own falsehood. We may therefore assume it to be false, in which case we have the contradiction that what it asserts is in fact true. Thus both the truth and falsehood of this statement are ruled out, and so it cannot be classed as a proposition. This example illustrates a rather nice distinction: to assert that a given statement is not true is not necessarily to assert that that statement is false (unless it is already known to be a proposition).

Such examples indicate that care is required in the formulation of propositions. It is essential that we avoid statements such as the one in the preceding example. We need to be more precise as to exactly which statements are to be deemed propositions. In the present chapter we attempt to achieve a satisfactory degree of precision in this regard. We also develop some rules for working with propositions.

It is worth noting that the statement considered in the previous example makes reference to itself. Such self-referential statements are discussed in detail in D. Hofstadter, *Gödel, Escher, Bach: An Eternal Golden Braid*, Vintage Books, New York, 1979.

The notion of a proposition is closely related to that of a collection of objects. Some such collections will be referred to as **sets**. The objects in a set X are called the **elements** or **members** of X, and are said to be **contained** in X or to **belong** to X. If an object x is an element of a set X, then we write $x \in X$; otherwise we write $x \notin X$. If a given collection is to be a set X, then for any object x we insist that the assertion of the containment of x in X must be a proposition. In other words, it must be possible in principle to determine whether or not x is an element of X, no matter what object x is. For instance, the collection of all flowers is a set which contains every rose, and the collection of all animals is a set containing no rose. The collection of all roses which are animals is a set containing nothing at all. The collection of all beautiful objects is not a set, because the assertion that a rose (or any other given object) is beautiful is, as we have seen, not a proposition.

As in the case of propositions, the question of whether a given collection is a set is sometimes a subtle one. For instance, consider the collection X of all false propositions. If that collection were a set, then the proposition "this statement is in X" would be the assertion of its own falsehood. But we have already seen that a statement which is the assertion of its own falsehood cannot in fact be a proposition. Therefore X cannot be a set.

Thus the concept of a set requires every bit as much care in its formulation as that of a proposition. We need a clear idea as to precisely which collections of objects are to be regarded as sets. We shall discuss this matter in more detail in Chapter 2. However a pertinent observation should be made here. In view of our earlier examples of sets, it seems reasonable to assert the existence of a set with no elements. This set is denoted by $\emptyset$. A set with no elements is said to be **empty** or **null**. The assertion that it contains a given object, such as itself, furnishes an example of a false proposition. Thus we can now assert the existence of propositions (albeit false ones). In this chapter we discuss, among other things, how new propositions (including true ones) can be constructed from those whose existence has already been established. A similar discussion for sets appears in Chapter 2.

Difficulties such as those we have touched on teased the minds of many mathematicians early in the twentieth century. Struggling with these problems, they were led to a clearer understanding of the concepts of a proposition and of a set. The clarification of these ideas in turn has led to a more secure underpinning for the whole of mathematics. For more details, consult H. Eves, C.V. Newsom, *An Introduction To The Foundations And Fundamental Concepts Of Mathematics*, Holt, Rinehart and Winston, New York, 1965.

1.2 Propositions

Let p be a proposition. The assertion that p is false is another proposition, called the **negation** of p. We shall denote the negation of p by $\sim p$. For example, if p is the proposition that a rose is an animal, then $\sim p$ is the proposition that a rose is not an animal. Note that $\sim p$ is true if and only if p is false. In other words, if $\sim p$ is true then p is false, and if p is false then $\sim p$ is true.

Let p and q be propositions. The assertion that either p or q is true is another proposition, called the **disjunction** of p and q. The word "or" is always to be interpreted in its inclusive sense. In other words, we admit the possibility that both p and q may be true. Indeed, the only possibility being ruled out is that p and q are both false. We denote the disjunction of p and q by $p \vee q$. For example, if p is the proposition that a rose is a flower and q the proposition that a rose is an animal, then $p \vee q$ is the proposition that a rose is a flower or an animal.

The operation of negation of a proposition has priority over that of disjunction of propositions and every operation yet to be defined. For example, if p is a proposition then

$$\sim p \vee p$$

denotes the proposition $(\sim p) \vee p$, not $\sim (p \vee p)$. Note that $\sim p \vee p$ is always true, since p is necessarily either true or false. Thus there exist true propositions, as well as false ones.

At this juncture we introduce the idea of a **truth table**. Consider the proposition $p \vee q$, where p and q are propositions. We construct a table with columns headed p, q and $p \vee q$. The rows give all the possible combinations of truth and falsehood for these propositions, truth being denoted by T and falsehood by F. The truth table for $p \vee q$ is given in Table 1.1. Similarly Table 1.2 gives the truth table for $\sim p$.

We have discussed two operations on propositions: the negation $\sim p$ of a proposition p and the disjunction $p \vee q$ of propositions p and q. By

Table 1.1

p	q	$p \vee q$
T	T	T
T	F	T
F	T	T
F	F	F

Table 1.2

p	$\sim p$
T	F
F	T

combining these operations we can construct new propositions. An example is the proposition

$$\sim (\sim p \vee \sim q),$$

which we denote by $p \wedge q$. This proposition is called the **conjunction** of p and q, and its truth table is constructed in Table 1.3. The truth table shows that $p \wedge q$ is true if and only if p and q are both true. For example, if p is the proposition that a rose is a flower and q the proposition that it is an animal, then $p \wedge q$ is the (false) proposition that a rose is both a flower and an animal.

Table 1.3

p	q	$\sim p$	$\sim q$	$\sim p \vee \sim q$	$p \wedge q$
T	T	F	F	F	T
T	F	F	T	T	F
F	T	T	F	T	F
F	F	T	T	T	F

For any proposition p we note that

$$\sim p \wedge p$$

is always false, since no proposition can be both true and false.

If p and q are propositions, we define $p \Rightarrow q$ to be the proposition

$$\sim p \vee q.$$

This proposition is called an **implication**. If it is true, then we say that p **implies** q. The truth table for $p \Rightarrow q$ is given in Table 1.4. Observe that if $p \Rightarrow q$ is true, then q must be true if p is true. In other words, any proposition implied by a true proposition must be true. However no conclusion can be drawn concerning q if p is false.

Table 1.4

p	q	$\sim p$	$p \Rightarrow q$
T	T	F	T
T	F	F	F
F	T	T	T
F	F	T	T

A proposition is said to **hold** if it is true. It is tempting to interpret the truth of the proposition $p \Rightarrow q$ as meaning that if p holds then so does q. Certainly if p and $p \Rightarrow q$ both hold then the truth table confirms that q holds as well. But there is a bit more to it than that: the truth table shows that $p \Rightarrow q$ also holds if p is false, no matter which proposition q is. We may then say that the implication $p \Rightarrow q$ is **vacuous**. Thus any proposition at all is implied by a false proposition. For example, the proposition that a rose is an animal implies that a pig can fly.

Similarly any proposition implies a given true proposition. Indeed, Table 1.4 shows that to confirm the truth of the proposition $p \Rightarrow q$ we need only rule out the possibility of p being true and q false. In other words it suffices to show that if p is true then so is q.

If p implies q, we may say that p is **sufficient** for q and that q is **necessary** for p. The latter terminology is motivated by the observation that if p implies q and q is false, then p must also be false by Table 1.4.

We sometimes refer to p as the **premise** or **hypothesis**, and to q as the **conclusion**, of the implication $p \Rightarrow q$.

If p and q are propositions, then the proposition $q \Rightarrow p$ is called the **converse** of $p \Rightarrow q$.

We draw attention to some elementary properties of this concept of implication. Firstly, it is clear that any proposition implies itself. Indeed, we have already seen that the proposition $\sim p \vee p$ holds for any proposition p.

Secondly, let p, q, r be propositions such that $p \Rightarrow q$ and $q \Rightarrow r$ both hold. If p is true, then q is true and so r is true. We conclude that $p \Rightarrow r$ also holds. In other words, if p implies q and q implies r, then p implies r. This property may be applied repeatedly to an arbitrarily long chain of implications. The premise of the first implication in the chain then implies the conclusion of the last one. For example, if p, q, r, s are propositions such that $p \Rightarrow q$, $q \Rightarrow r$ and $r \Rightarrow s$ all hold, then $p \Rightarrow r$ holds, and since $r \Rightarrow s$ also holds we conclude that $p \Rightarrow s$ holds. The construction of such chains of implications forms the essence of mathematical proofs. We deem a proposition q to have been proved, and to be a **theorem**, once we have found a proposition p such that

$$p \wedge (p \Rightarrow q)$$

is known to be true, either by hypothesis or by a previous proof. In other words, a proposition is a theorem if it is implied by some proposition which has previously been hypothesized or proved. For example, $\sim p \vee p$ is a theorem for any proposition p, since it is true and (like any proposition) implies itself. Note that every theorem is true, being implied by a true proposition.

Let p, q, r be propositions. The proposition

$$(p \Rightarrow q) \wedge (q \Rightarrow r)$$

is usually written more briefly as

$$p \Rightarrow q \Rightarrow r.$$

If it holds then, as we have already seen, $p \Rightarrow r$ also holds.

Suppose now that p is a true proposition and that the implication

$$\sim q \Rightarrow \sim p$$

holds for some proposition q. Since $\sim p$ is false and necessary for $\sim q$, it follows that $\sim q$ is false, so that q is true. We have therefore shown that the proposition

$$(\sim q \Rightarrow \sim p) \Rightarrow (p \Rightarrow q)$$

holds for any propositions p and q. Thus in order to prove the proposition $p \Rightarrow q$, where p and q are any propositions, it suffices to prove that

$$\sim q \Rightarrow \sim p$$

holds. This proposition is called the **contrapositive** of $p \Rightarrow q$.

The considerations of the previous paragraph show that in order to prove a proposition q it suffices to find a true proposition p such that $\sim q$ implies $\sim p$. This method of proving q is referred to as **proof by contradiction**.

It should not be supposed that every true statement has a proof. For example, consider the following statement: "this statement has no proof". If this statement does have a proof, then the existence of that proof makes the statement false. We therefore have a contradiction, since no false statement can possibly have a proof and hence be a theorem. We conclude that the statement in question has no proof, and is therefore true. For any proposition p either p or $\sim p$ must be true, but it may be that neither has a proof. In this case we say that p is **undecidable**.

For propositions p and q we define the proposition $p \Leftrightarrow q$ to be

$$p \Rightarrow q \Rightarrow p.$$

This proposition is called an **equivalence**. Its truth table is constructed in Table 1.5. If $p \Leftrightarrow q$ holds, then we say that p and q are **equivalent** (to each other).

Table 1.5

p	q	$p \Rightarrow q$	$q \Rightarrow p$	$p \Leftrightarrow q$
T	T	T	T	T
T	F	F	T	F
F	T	T	F	F
F	F	T	T	T

From the truth table, we see that p and q are equivalent if and only if both hold or neither does. In other words, to say that p and q are equivalent is to say that p holds if and only if q holds. Yet another way of expressing this condition is to assert that p is both necessary and sufficient for q. Alternatively, we can say that p and q imply each other.

It should be evident that any proposition is equivalent to itself, and that if proposition p is equivalent to proposition q then q is also equivalent to p. Suppose p, q, r are propositions such that $p \Leftrightarrow q$ and $q \Leftrightarrow r$ both hold. Then $p \Rightarrow q \Rightarrow r$ and $r \Rightarrow q \Rightarrow p$ hold, and therefore so does $p \Leftrightarrow r$. The proposition

$$(p \Leftrightarrow q) \wedge (q \Leftrightarrow r)$$

is usually shortened to

$$p \Leftrightarrow q \Leftrightarrow r.$$

Thus we have shown that if $p \Leftrightarrow q \Leftrightarrow r$ holds then $p \Leftrightarrow r$ holds also. This property may be applied to arbitrarily long chains of equivalences.

If p and q are propositions, then a comparison of the appropriate truth tables shows that $p \vee q$ and $q \vee p$ are equivalent, as are $p \wedge q$ and $q \wedge p$. (For example, for the former equivalence it suffices to note that in Table 1.1 the entries for $p \vee q$ are the same in the rows where the entries for p and q differ. A similar inspection of Table 1.3 verifies the latter equivalence.) Similarly $p \vee p$ and $p \wedge p$ are both equivalent to p.

If p and r are equivalent propositions, then certainly

$$\sim p \Leftrightarrow \sim r$$

holds. Moreover Table 1.6 shows that

$$(p \vee q) \Leftrightarrow (r \vee q)$$

holds: just compare the entries for $p \vee q$ and $r \vee q$ in the rows where the entries for p and r agree. It follows that

$$q \vee p \Leftrightarrow p \vee q \Leftrightarrow r \vee q \Leftrightarrow q \vee r.$$

Therefore, given one of the propositions $\sim p$, $p \vee q$ and $q \vee p$, any proposition equivalent to p may be substituted for p without affecting the veracity of the given proposition. Clearly an analogous result holds for any proposition formed from p and other propositions by any combination of the negation and disjunction operations. For example, if p and r are equivalent propositions then

$$(p \wedge q) \Leftrightarrow (r \wedge q),$$
$$(p \Rightarrow q) \Leftrightarrow (r \Rightarrow q),$$
$$(q \Rightarrow p) \Leftrightarrow (q \Rightarrow r),$$
$$(p \Leftrightarrow q) \Leftrightarrow (r \Leftrightarrow q)$$

all hold for any proposition q.

Table 1.6

p	q	r	$p \vee q$	$r \vee q$
T	T	T	T	T
T	T	F	T	T
T	F	T	T	T
T	F	F	T	F
F	T	T	T	T
F	T	F	T	T
F	F	T	F	T
F	F	F	F	F

Let us continue to augment our stock of equivalences. Table 1.7 reveals the equivalence

$$\sim\sim p \Leftrightarrow p$$

for any proposition p. We refer to this equivalence as the **principle of double negation**.

Table 1.7

p	$\sim p$	$\sim\sim p$
T	F	T
F	T	F

For example, since a rose is a flower it is false that a rose is not a flower. Conversely, since it is false that a rose is not a flower, it is indeed a flower.

As an application, if p and q are propositions then

$$p \vee q \Leftrightarrow \sim\sim p \vee q$$
$$\Leftrightarrow (\sim p \Rightarrow q).$$

Now suppose that q implies another proposition r. It follows that the implication

$$(p \vee q) \Rightarrow (p \vee r)$$

holds, for if $p \vee q$ holds then

$$\sim p \Rightarrow q \Rightarrow r$$

so that $p \vee r$ holds. If p also implies r, then by using the same argument we obtain

$$p \vee q \Rightarrow p \vee r \Rightarrow r \vee p \Rightarrow r \vee r \Rightarrow r.$$

In other words, if $p \vee q$ holds and p and q both imply r then r must also hold. Thus in order to prove a proposition r it suffices to find propositions p and q such that $p \vee q$, $p \Rightarrow r$ and $q \Rightarrow r$ all hold. This method of proving r is sometimes called **disjunction of cases**.

In Table 1.8 we construct the truth table for the proposition

$$\sim (\sim p \wedge \sim q)$$

where p and q are any propositions. By comparing Table 1.8 with Table 1.1 we infer that not only does

$$(p \wedge q) \Leftrightarrow \sim (\sim p \vee \sim q)$$

hold (as we already knew), but

$$(p \vee q) \Leftrightarrow \sim (\sim p \wedge \sim q)$$

holds as well. Each of these equivalences may be obtained formally from the other by interchanging the symbols $\vee$ and $\wedge$. We express this relationship by declaring these equivalences to be **dual** to each other.

Table 1.8

p	q	$\sim p$	$\sim q$	$\sim p \wedge \sim q$	$\sim(\sim p \wedge \sim q)$
T	T	F	F	F	T
T	F	F	T	F	T
F	T	T	F	F	T
F	F	T	T	T	F

By using substitution and the principle of double negation, we verify the following equivalences:

$$\sim (p \wedge q) \Leftrightarrow \sim\sim (\sim p \vee \sim q)$$
$$\Leftrightarrow \sim p \vee \sim q.$$

The equivalence

$$\sim (p \wedge q) \Leftrightarrow (\sim p \vee \sim q)$$

and its dual,

$$\sim (p \vee q) \Leftrightarrow (\sim p \wedge \sim q),$$

which clearly holds by the same argument, are called **de Morgan's laws**.

In Table 1.9 and Table 1.10 we construct the truth tables for the propositions

$$p \wedge (q \vee r)$$

and

$$(p \wedge q) \vee (p \wedge r)$$

respectively, where p, q, r are any propositions. Comparison of the truth tables reveals the equivalence of these two propositions.

We also verify the dual of this equivalence, either by another comparison of the relevant truth tables or by using equivalences we have already verified, such as the principle of double negation and those involving the negations of disjunctions and conjunctions, to check that all the equivalences in the

Table 1.9

p	q	r	q ∨ r	p ∧ (q ∨ r)
T	T	T	T	T
T	T	F	T	T
T	F	T	T	T
T	F	F	F	F
F	T	T	T	F
F	T	F	T	F
F	F	T	T	F
F	F	F	F	F

Table 1.10

p	q	r	p ∧ q	p ∧ r	(p ∧ q) ∨ (p ∧ r)
T	T	T	T	T	T
T	T	F	T	F	T
T	F	T	F	T	T
T	F	F	F	F	F
F	T	T	F	F	F
F	T	F	F	F	F
F	F	T	F	F	F
F	F	F	F	F	F

following chain hold:

$$\begin{aligned} p \vee (q \wedge r) &\Leftrightarrow \sim (\sim p \wedge \sim (q \wedge r)) \\ &\Leftrightarrow \sim (\sim p \wedge (\sim q \vee \sim r)) \\ &\Leftrightarrow \sim ((\sim p \wedge \sim q) \vee (\sim p \wedge \sim r)) \\ &\Leftrightarrow \sim (\sim (p \vee q) \vee \sim (p \vee r)) \\ &\Leftrightarrow (p \vee q) \wedge (p \vee r). \end{aligned}$$

We have already seen that $\sim p \vee p$ holds for any proposition p. Thus the equivalence

$$(\sim p \vee p) \Leftrightarrow q$$

holds for any true proposition q. In this chapter we shall use the symbol 1 for a particular true proposition and the symbol 0 for a particular false one. Thus

$$\sim 0 \Leftrightarrow 1$$

and

$$\sim 1 \Leftrightarrow 0$$

hold. With this notation, we see that the equivalence

$$(\sim p \vee p) \Leftrightarrow 1$$

also holds. Similarly

$$(\sim p \wedge p) \Leftrightarrow 0$$

holds, since $\sim p \wedge p$ is false. (This equivalence can also be obtained from the previous one, using the definition of conjunction.) Note that these equivalences can be incorporated into the framework of duality if we agree to interchange 0 and 1 as well as $\vee$ and $\wedge$.

By substituting 0 for q in Table 1.1, we find that the equivalence

$$(p \vee 0) \Leftrightarrow p$$

holds for any proposition p since q is then false. Dually we verify the equivalence

$$(p \wedge 1) \Leftrightarrow p$$

by substituting 1 for q in Table 1.3. Similarly the equivalences

$$(p \vee 1) \Leftrightarrow 1$$

and

$$(p \wedge 0) \Leftrightarrow 0$$

can be confirmed.

In summary, we list some pairs of dual equivalences we have now established. For ease of reference, this result is stated as a theorem.

Theorem 1.1 *Let p, q, r be propositions. Then the following equivalences hold:*

(a) $(p \vee q) \Leftrightarrow (q \vee p)$ *and* $(p \wedge q) \Leftrightarrow (q \wedge p)$;

(b) $(p \wedge (q \vee r)) \Leftrightarrow ((p \wedge q) \vee (p \wedge r))$ and $(p \vee (q \wedge r)) \Leftrightarrow ((p \vee q) \wedge (p \vee r))$;
(c) $(p \wedge \sim p) \Leftrightarrow 0$ and $(p \vee \sim p) \Leftrightarrow 1$;
(d) $(p \vee 0) \Leftrightarrow p$ and $(p \wedge 1) \Leftrightarrow p$.

Property (a) above may be described by saying that the operations of disjunction and conjunction are **commutative** (with respect to equivalence), and that p and q **commute** under disjunction and conjunction. We may refer to the equivalences in (a) as the **commutative laws**. Similarly the first equivalence in (b) is described by the assertion that conjunction is **distributive** over disjunction (with respect to equivalence). By the second equivalence in (b), disjunction is also distributive over conjunction. These equivalences are called the **distributive laws**. The equivalences in (d) are summarized by describing 0 and 1 as **identities** (with respect to equivalence) under disjunction and conjunction respectively. (In fact, any false proposition is an identity under disjunction while any true proposition is an identity under conjunction.)

It turns out that all the other equivalences we have established may be obtained from those listed in Theorem 1.1, without reference to truth tables. We list some of these equivalences below, where p is any proposition:

$$\sim\sim p \Leftrightarrow p, \qquad (1.1)$$

$$\sim 0 \Leftrightarrow 1, \qquad (1.2)$$

$$\sim 1 \Leftrightarrow 0, \qquad (1.3)$$

$$(p \vee p) \Leftrightarrow (p \wedge p) \Leftrightarrow p, \qquad (1.4)$$

$$(p \vee 1) \Leftrightarrow 1, \qquad (1.5)$$

$$(p \wedge 0) \Leftrightarrow 0. \qquad (1.6)$$

In addition, Theorem 1.1 and the results listed above can be used to verify additional equivalences. An example is supplied by the next theorem. Note first that the dual of each equivalence in Theorem 1.1 or the above list is also true. Therefore the dual of any result proved from these equivalences must also hold.

Theorem 1.2 *For any propositions p and q, the equivalences*

$$((q \wedge p) \vee p) \Leftrightarrow ((q \vee p) \wedge p) \Leftrightarrow p$$

hold.

Proof. The following equivalences hold:

$$(q \vee p) \wedge p \Leftrightarrow (p \vee q) \wedge (p \vee 0) \quad \text{(by Theorem 1.1(a),(d))}$$
$$\Leftrightarrow p \vee (q \wedge 0) \quad \text{(by Theorem 1.1(b))}$$
$$\Leftrightarrow p \vee 0 \quad \text{(by 1.6)}$$
$$\Leftrightarrow p \quad \text{(by Theorem 1.1(d)).}$$

Thus the second of the required equivalences holds, and the rest follows by duality. □

The equivalences in Theorem 1.2 are sometimes called the **absorption laws**.

Before giving another example, we present a lemma.

Lemma 1.3 *Let p, q, r be propositions. If*

$$(p \vee r) \Leftrightarrow (q \vee r)$$

and

$$(p \vee \sim r) \Leftrightarrow (q \vee \sim r)$$

both hold, then p and q are equivalent.

Proof. Since the equivalences $(p \vee r) \Leftrightarrow (q \vee r)$ and $(p \vee \sim r) \Leftrightarrow (q \vee \sim r)$ both hold, so does

$$((p \vee r) \wedge (p \vee \sim r)) \Leftrightarrow ((q \vee r) \wedge (q \vee \sim r)) \qquad (1.7)$$

by substitution. Moreover the following equivalences all hold (though it is left to the reader to determine which results have been used):

$$(p \vee r) \wedge (p \vee \sim r) \Leftrightarrow p \vee (r \wedge \sim r)$$
$$\Leftrightarrow p \vee 0$$
$$\Leftrightarrow p.$$

By symmetry the equivalence

$$((q \vee r) \wedge (q \vee \sim r)) \Leftrightarrow q$$

also holds. The truth of the equivalence $p \Leftrightarrow q$ now follows by substitution into 1.7. □

Theorem 1.4 *Let p, q, r be propositions. Then the following equivalences hold:*

$$(p \vee (q \vee r)) \Leftrightarrow ((p \vee q) \vee r), \tag{1.8}$$

$$(p \wedge (q \wedge r)) \Leftrightarrow ((p \wedge q) \wedge r). \tag{1.9}$$

Proof. The second equivalence follows by verifying the equivalences in the chains

$$\begin{aligned}(p \wedge (q \wedge r)) \vee r &\Leftrightarrow (p \vee r) \wedge ((q \wedge r) \vee r) \\ &\Leftrightarrow (p \vee r) \wedge r \\ &\Leftrightarrow r \\ &\Leftrightarrow ((p \wedge q) \wedge r) \vee r\end{aligned}$$

and

$$\begin{aligned}(p \wedge (q \wedge r)) \vee \sim r &\Leftrightarrow (p \vee \sim r) \wedge ((q \wedge r) \vee \sim r) \\ &\Leftrightarrow (p \vee \sim r) \wedge ((q \vee \sim r) \wedge (r \vee \sim r)) \\ &\Leftrightarrow (p \vee \sim r) \wedge ((q \vee \sim r) \wedge 1) \\ &\Leftrightarrow (p \vee \sim r) \wedge (q \vee \sim r) \\ &\Leftrightarrow (p \wedge q) \vee \sim r \\ &\Leftrightarrow ((p \wedge q) \vee \sim r) \wedge 1 \\ &\Leftrightarrow ((p \wedge q) \vee \sim r) \wedge (r \vee \sim r) \\ &\Leftrightarrow ((p \wedge q) \wedge r) \vee \sim r\end{aligned}$$

and appealing to Lemma 1.3. The first equivalence follows by duality. □

Theorem 1.4 may be expressed by describing the operations of disjunction and conjunction as **associative** (with respect to equivalence). The equivalences 1.8 and 1.9 are referred to as the **associative laws**. In view of 1.8 we often write $p \vee q \vee r$ instead of $p \vee (q \vee r)$ or $(p \vee q) \vee r$, where p, q, r are propositions. Similarly we may write $p \wedge q \wedge r$ in place of $p \wedge (q \wedge r)$ or $(p \wedge q) \wedge r$.

1.3 Predicates and Quantifiers

Let p be a statement about an unspecified object x, which will be called a **variable**. Then p might not be a proposition. For example, the statement "x is a rose" is not a proposition because its truth cannot be confirmed or denied until the identity of x is revealed. However if $p(x_0)$ denotes the statement obtained from p by substituting a particular object x_0 for the variable x, then it may be that $p(x_0)$ is a proposition for each choice of x_0.

In this case p is called a **predicate**. We may also say that p **depends** on x. The previous example is a predicate depending on x, because as soon as the object x is given to us we can always ascertain whether or not it is a rose.

If p is a predicate depending on a variable x and $p(x_0)$ holds for a particular choice x_0 of x, then we may say that p **holds** for x_0, and that x_0 **satisfies** p.

Given a predicate p depending on a variable x, it is either true or false that there exists a choice x_0 of x for which $p(x_0)$ holds. The assertion of the existence of such an x_0 is therefore a proposition. This proposition is denoted by $(\exists x)p$, and the symbol $\exists$ is called the **existential quantifier**. For example, if p is the predicate that x is a rose, then $(\exists x)p$ is the proposition that a rose exists. The proposition $(\exists x)p$ may be proved by producing an example of an x for which the predicate p is true.

Observe that the truth of the proposition $(\exists x)p$ does not depend on a choice of x. Thus $(\exists x)p$ is equivalent to $(\exists y)p'$ for any variable y, where p' is the predicate obtained from p by the substitution of y for x. For instance, the proposition "a rose exists" is equivalent to both "there exists x such that x is a rose" and "there exists y such that y is a rose". For this reason x may be referred to as a **dummy variable** in the proposition $(\exists x)p$.

The proposition

$$\sim (\exists x)(\sim p)$$

asserts that there is no choice of x for which p is false. In other words, the predicate p is true for each x. This proposition will be written as $(\forall x)p$, the symbol $\forall$ being called the **universal quantifier**. Again, x is a dummy variable. For example, if p is the predicate that x is a rose, then $(\forall x)p$ denotes the proposition that everything is a rose. The proposition $(\forall x)p$ may be proved by showing that the predicate p is true for an arbitrarily chosen x.

Note that the proposition

$$\sim (\forall x)(\sim p)$$

is the negation of the assertion that p is false for each x. It is therefore equivalent to $(\exists x)p$.

From the definition of $(\forall x)p$ and the principle of double negation, we find that the proposition

$$\sim (\forall x)p \Leftrightarrow (\exists x)(\sim p)$$

holds. Similarly we verify

$$\sim (\exists x)p \Leftrightarrow (\forall x)(\sim p).$$

These equivalences may be regarded as extensions of de Morgan's laws.

We have now introduced all the basic kinds of propositions that we shall need. They are the assertion of the membership of a given object in a given set, the negation of a proposition, the disjunction of propositions and the assertion of the existence of an object satisfying a given predicate. Accordingly, from this point onwards it is only propositions of these types that we shall contemplate. Any other assertion will henceforth be described simply as a statement, even if we can establish it to be true or false.

Earlier we showed that the collection of all false propositions is not a set. The same argument shows that this statement remains valid even with our new, more restrictive interpretation of the word "proposition".

Before we conclude our discussion of propositions, let us check that "this statement is false" is not a statement of a type that we have been discussing, and therefore has not been inadvertently declared to be a proposition. First, it is not an assertion of membership in a set, since the collection of all false statements is not a set. It is not the negation of any proposition. (It is not the negation of the statement "this statement is true", as it is in fact an assertion about another statement.) It is clearly not a disjunction of propositions different from itself. Finally, it is not the assertion of the existence of a false statement (even though it purports to furnish an example of one), for it is not implied by the existence of a false statement.

Exercises 1

(1) Determine whether the following statements are true, false or neither:
   (a) "This statement is false" is true.
   (b) "This statement is false" is not true.
   (c) "This statement is false" is false.
   (d) "This statement is false" is not false.

(2) Let p and q be propositions. Construct the truth table for the proposition

$$(p \vee q) \wedge \sim (p \wedge q).$$

(3) For any propositions p and q, denote by $p + q$ the proposition given in

Exercise 2. Let r, s, t be propositions. Show that the propositions
$$r \wedge (s + t)$$
and
$$(r \wedge s) + (r \wedge t)$$
are equivalent.

(4) Let p and q be propositions such that $(p \vee q) \Leftrightarrow 1$ and $(p \wedge q) \Leftrightarrow 0$ hold. Prove from Theorem 1.1 and its consequences that q is equivalent to $\sim p$.

(5) Prove the following equivalences from Theorem 1.1 and its consequences:

(a) $0 \Leftrightarrow p$, where p is any false proposition;
(b) $1 \Leftrightarrow p$, where p is any true proposition;
(c) $\sim\sim p \Leftrightarrow p$ for any proposition p;
(d) $\sim 0 \Leftrightarrow 1$ and $\sim 1 \Leftrightarrow 0$;
(e) $(p \vee p) \Leftrightarrow (p \wedge p) \Leftrightarrow p$ for any proposition p;
(f) $(p \vee 1) \Leftrightarrow 1$ and $(p \wedge 0) \Leftrightarrow 0$ for any proposition p;
(g) de Morgan's laws. (Hint: Use Exercise 4.)

(6) Let p and q be propositions. Prove from Theorem 1.1 and its consequences that the propositions
$$(p \wedge q) \Leftrightarrow p,$$
$$(p \vee q) \Leftrightarrow q,$$
$$(p \wedge \sim q) \Leftrightarrow 0$$
are equivalent.

(7) Let $p|q$ denote the proposition
$$\sim (p \wedge q),$$
where p and q are propositions. Show that the equivalences
$$\sim r \Leftrightarrow (r|r)$$
and
$$(r \vee s) \Leftrightarrow ((r|r)|(s|s))$$
hold for any propositions r and s, and find corresponding equivalences for the propositions $r \wedge s$ and $r \Rightarrow s$.

(8) Let p, q, r be propositions. Prove that if
$$(p \vee r) \Leftrightarrow (q \vee r)$$
and
$$(p \wedge r) \Leftrightarrow (q \wedge r)$$
hold then p and q are equivalent.

Chapter 2

Sets

2.1 Construction of Sets

In Chapter 1 we asserted the existence of certain propositions and devised ways of constructing new propositions from them. We also asserted the existence of a set $\emptyset$ with no elements. In this chapter we proceed to construct new sets in a manner analogous to our construction of propositions.

First, however, let us make some definitions. Let X and Y be sets. (For example, X and Y could both be $\emptyset$.) We say that X is a **subset** of Y if each element of X is contained in Y. In other words, for any object x the proposition $x \in X$ implies that $x \in Y$. If X is a subset of Y, then we write $X \subseteq Y$; otherwise we write $X \nsubseteq Y$. The proposition $X \subseteq Y$ is called an **inclusion**. We say that a set **includes** its subsets. However, when we say that an object is in a set X, we will mean that it is contained in X rather than included in X.

Clearly any set includes itself. The following result is also easily demonstrated.

Theorem 2.1 *If X, Y, Z are sets such that $X \subseteq Y$ and $Y \subseteq Z$, then $X \subseteq Z$.*

Proof. We need to show that for any object x the implication

$$(x \in X) \Rightarrow (x \in Z)$$

holds. But $(x \in X) \Rightarrow (x \in Y)$ holds since $X \subseteq Y$, and $(x \in Y) \Rightarrow (x \in Z)$ holds since $Y \subseteq Z$. The required implication follows. $\square$

If X, Y, Z are sets such that $X \subseteq Y$ and $Y \subseteq Z$, then we usually write $X \subseteq Y \subseteq Z$. Theorem 2.1 shows that if $X \subseteq Y \subseteq Z$ then $X \subseteq Z$. This property may be applied to arbitrarily long chains of inclusions.

If X and Y are sets such that $X \subseteq Y \subseteq X$, then we say that sets X and Y are **equal** (to each other). Otherwise X and Y are said to be **distinct** (from each other). Intuitively, sets X and Y are equal if and only if they have the same elements. In this case we write $X = Y$; otherwise we write $X \neq Y$. The proposition $X = Y$ is called an **equation**.

Note that $X = Y$ if and only if the propositions $x \in X$ and $x \in Y$ are equivalent for each object x. Thus if $X = Y$ then $Y = X$. Moreover $X = X$ for any set X.

If X and Y are equal sets and x is any object, then $x \in X$ if and only if $x \in Y$. Since a set is determined by its elements, it therefore makes sense to regard X and Y as identical for all mathematical purposes. More precisely, let p be a predicate depending on a variable x. Recall that $p(X)$ and $p(Y)$ are the propositions obtained from p by the substitution of X and Y, respectively, for x. Since $X = Y$, it follows from the identification of X with Y that $p(X)$ is equivalent to $p(Y)$. In other words, X may be substituted for Y, or vice versa, without affecting the veracity of the proposition in question. For example, let X, Y, Z be any sets. If $X = Y$ and $Y \subseteq Z$, then it follows that $X \subseteq Z$. (Here we took p to be the predicate $x \subseteq Z$, which depends on the variable x.) Similarly if $X \subseteq Y$ and $Y = Z$ then again $X \subseteq Z$, and if $X = Y$ and $Y = Z$ then $X = Z$.

Note that our agreement to regard equal sets X and Y as identical amounts to the assumption that if $X \in Z$ for some set Z then $Y \in Z$ also.

The following theorem summarizes some of the important results concerning equality that we have already established.

Theorem 2.2 *Let X, Y, Z be sets.*

(a) $X = X$.
(b) *If $X = Y$ then $Y = X$.*
(c) *If $X = Y$ and $Y = Z$ then $X = Z$.*

If $X = Y$ and $Y = Z$, then we usually write $X = Y = Z$ instead. If $X = Y = Z$ then Theorem 2.2(c) shows that $X = Z$. This property may be applied to arbitrarily long chains of equations.

Let X and Y be sets. If $X \subseteq Y$ and $X \neq Y$, then we say that X is a **proper** subset of Y. In this case we write $X \subset Y$; otherwise we write $X \not\subset Y$. The proposition $X \subset Y$ is called a **proper** inclusion. Note that $X \subseteq Y$ if and only if $X \subset Y$ or $X = Y$. Indeed, this result is clear if $X \subset Y$ or $X = Y$. Suppose therefore that $X \subseteq Y$. If $X \neq Y$ then $X \subset Y$ by definition. We conclude that either $X = Y$ or $X \subset Y$.

Theorem 2.3 *Let X, Y, Z be sets. If $X \subset Y$ and $Y \subset Z$, then $X \subset Z$.*

Proof. Since $X \subset Y$ and $Y \subset Z$ it follows that $X \subseteq Y \subseteq Z$. It remains only to show that $X \neq Z$. Since $X \subseteq Y$ and $X \neq Y$, we must have $Y \nsubseteq X$. Therefore there exists $y \in Y$ such that $y \notin X$. Since $Y \subseteq Z$ it follows that $y \in Z$. Since $y \in Z$ and $y \notin X$ we deduce that $Z \nsubseteq X$, and so $Z \neq X$, as required. □

If $X \subset Y$ and $Y \subset Z$, then we usually write $X \subset Y \subset Z$ instead. If this condition holds, then Theorem 2.3 shows that $X \subset Z$. This result may be applied to arbitrarily long chains of proper inclusions.

If either $X \subset Y$ and $Y \subseteq Z$, or $X \subseteq Y$ and $Y \subset Z$, then $X \subset Z$. Indeed, this conclusion is immediate if $X = Y$ or $Y = Z$, and follows from Theorem 2.3 in the remaining case.

At this point we introduce the idea of a **Venn diagram**, in which a set is represented by a region enclosed by a non-intersecting closed curve drawn in the plane. If X and Y are sets such that $X \subseteq Y$, then the region representing X is to be contained inside the region representing Y. Of course equal sets are to be represented by the same region, and distinct sets by different regions. By means of this device, results such as Theorem 2.3 become evident geometrically. (See Figure 2.1.) Venn diagrams are somewhat akin to truth tables.

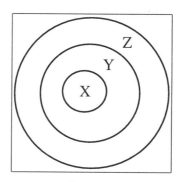

Fig. 2.1

Let us now prove some theorems about an empty set X. We shall show that in fact $X = \emptyset$, but first we prove the following theorem.

Theorem 2.4 *Let X and Y be sets. If X is empty, then $X \subseteq Y$.*

Proof. We need to show that the implication $(x \in X) \Rightarrow (x \in Y)$ holds for any object x. But this implication is immediate, since its premise is false. □

Theorem 2.5 *If X is an empty set, then $X = \emptyset$.*

Proof. Since both X and $\emptyset$ are empty, we have $\emptyset \subseteq X \subseteq \emptyset$ by Theorem 2.4. Hence $X = \emptyset$. □

Let p be a predicate, and suppose that $X = Y$ whenever X and Y are sets satisfying p. We then say that X is the **unique** set which satisfies p. Theorem 2.5 thus shows that $\emptyset$ is the only empty set. Alternatively, we may express Theorem 2.5 by saying that the empty set is unique.

Theorem 2.6 *If X is a subset of $\emptyset$, then $X = \emptyset$.*

Proof. Suppose $X \neq \emptyset$. By Theorem 2.5 it follows that X is not empty. Therefore there exists $x \in X$. Since $x \notin \emptyset$, we reach the contradiction that $X \not\subseteq \emptyset$. Therefore $X = \emptyset$. □

Thus we can now conclude that $\emptyset$ has a unique subset. Indeed, it has $\emptyset$ as a subset, and if $X \subseteq \emptyset$ and $Y \subseteq \emptyset$ then $X = \emptyset = Y$ by Theorem 2.6, so that $\emptyset$ is the only subset of $\emptyset$.

So far $\emptyset$ is the only set whose existence has been guaranteed. If $\emptyset$ were the only set, then it would seem foolish to have developed the theory presented so far in this chapter. However this is not the case. We proceed to consider ways of constructing new sets from existing sets.

Let X be a set. We assert the existence of a set whose elements are the subsets of X. This set is called the **power set** of X, and is denoted by $\mathcal{P}(X)$. As a particular case, we find from Theorem 2.6 that $\mathcal{P}(\emptyset)$ is a set whose only element is $\emptyset$. This set is denoted by $\{\emptyset\}$. Note that it is distinct from $\emptyset$: it contains an element whereas $\emptyset$ does not.

The only subsets of $\{\emptyset\}$ are itself and $\emptyset$. Thus $\{\emptyset\}$ and $\emptyset$ are the only elements of $\mathcal{P}(\{\emptyset\})$. We denote this set by $\{\emptyset, \{\emptyset\}\}$. Let us now agree that if X is any set, then the replacement of the elements of X by sets yields another set. For example, let X be any set. If we replace the element $\emptyset$ of the set $\{\emptyset\}$ by X, then we obtain a set whose only element is X. This set is denoted by $\{X\}$, and is sometimes referred to as a **singleton**. Similarly, let X and Y be sets. If we replace the elements $\emptyset$ and $\{\emptyset\}$ of the set $\{\emptyset, \{\emptyset\}\}$ by X and Y respectively, then we obtain a set whose only elements are X and Y. This set is denoted by $\{X, Y\}$. Note that $\{X, Y\} = \{Y, X\}$, and that $\{X, Y\} = \{X\}$ if $X = Y$.

Let X be a set and p a predicate depending on a variable x. For any choice x_0 of x, we can form the proposition

$$p(x_0) \wedge (x_0 \in X).$$

The collection of all objects x_0 for which this proposition holds is then a set Y, which may be denoted by $\{x \in X : p\}$. It is a subset of X, and may be described as the set of elements of X for which p holds. The predicate p may be thought of as expressing a property which may hold for some choices of the variable x. In the formation of the set Y described above, these choices are to be made from X.

The reader may wonder why X is hypothesized to be a set in the definition of Y above. Suppose, for example, that X were the collection of all false propositions. If p were the predicate that the variable x is a false proposition, then $p(x_0)$ would hold for each x_0 in X. The set Y, as defined above, would therefore be identical to X. This result would contradict the fact that the collection of all false propositions is not a set.

Let X and Y be sets and let p be the predicate that $x \in Y$, where x is a variable. Then the set $\{x \in X : p\}$ is called the **intersection** of X and Y, and is denoted by $X \cap Y$. Its elements are the objects contained in both X and Y. For example, $\{X, Y\} \cap \{X\} = \{X\}$, and $\{X\} \cap \{Y\} = \emptyset$ if $X \neq Y$. Moreover if Z is a set distinct from X then $\{X, Y\} \cap \{Y, Z\} = \{Y\}$. We say that sets X and Y are **disjoint** (from each other) if $X \cap Y = \emptyset$, and that they **meet** (each other) otherwise. For instance $\{X, Y\}$ and $\{X\}$ meet, but if $X \neq Y$ then $\{X\}$ and $\{Y\}$ are disjoint.

Let X be a set whose elements are sets. Suppose that Y and Z are disjoint sets whenever Y and Z are distinct elements of X. Then we say that the elements of X are **mutually disjoint**.

Clearly

$$X \cap Y = Y \cap X$$

for any sets X and Y. In other words, the intersection operation is commutative. (The properties of commutativity, associativity and distributivity for operations on sets are always understood to be with respect to equality.) Indeed, this property follows immediately from the definition of intersection and the commutativity of conjunction of propositions. Similarly we find that

$$X \cap X = X$$

and
$$X \cap \emptyset = \emptyset$$

for any set X. (These equations follow from the fact that the equivalences $(p \wedge p) \Leftrightarrow p$ and $(p \wedge 0) \Leftrightarrow 0$ hold for any proposition p.) Note also that

$$X \cap Y \subseteq X$$

and

$$X \cap Y \subseteq Y$$

for any sets X and Y, and that

$$X \cap Y = X$$

if $X \subseteq Y$.

The intersection of sets X and Y may easily be visualized by using a Venn diagram. The region representing $X \cap Y$ is the overlap between the region representing X and that representing Y. It is indicated by the shaded area in Figure 2.2.

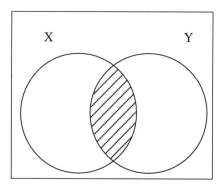

Fig. 2.2

Again, let X and Y be sets and let p be the predicate that $x \notin Y$, where x is a variable. Then the set $\{x \in X : p\}$ is the **complement** of Y with respect to X, and is denoted by $X - Y$. Its elements are those of X which are not in Y. For example, $\{X, Y\} - \{X\} = \{Y\}$ if $X \neq Y$. In a Venn diagram, $X - Y$ is represented by the shaded region in Figure 2.3.

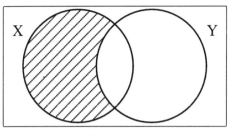

Fig. 2.3

Since the equivalence $(p \wedge \sim p) \Leftrightarrow 0$ holds for any proposition p, we have

$$Y \cap (X - Y) = \emptyset$$

for any sets X and Y. Similarly if $Y \subseteq X$ the equation

$$X - (X - Y) = Y$$

follows from the principle of double negation. We note also that

$$X - \emptyset = X$$

and

$$X - X = \emptyset$$

for any set X.

Now let X be a set whose elements are sets. We assert the existence of a set Y whose elements are the objects which are in a set in X. In other words, an object y is an element of Y if and only if there exists a set $Z \in X$ such that $y \in Z$. The set Y is called the **union** of X, and is denoted by $\cup X$. For example, $\cup \{\emptyset, \{\emptyset\}\} = \{\emptyset\}$, since $\emptyset$ has no elements and the only element of $\{\emptyset\}$ is $\emptyset$. For any sets X and Y we have $\cup\{\{X\},\{Y\}\} = \{X,Y\}$, which is equal to $\{X\}$ if $X = Y$. Note that $\cup \emptyset = \emptyset$.

If X and Y are sets, we define

$$X \cup Y = \cup\{X, Y\}.$$

(For example, $\{X\} \cup \{Y\} = \{X, Y\}$.) This set is called the **union** of X and Y. Its elements are the objects which are in X or Y. Thus

$$X \subseteq X \cup Y$$

and
$$Y \subseteq X \cup Y.$$

In a Venn diagram, $X \cup Y$ is represented by the shaded region in Figure 2.4.

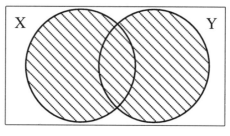

Fig. 2.4

The commutativity of disjunction of propositions immediately implies the commutativity of the union operation. A similar argument establishes the equations

$$X \cup \emptyset = X$$

and

$$X \cup X = X$$

for any set X. Similarly if $X \subseteq Y$ then

$$X \cup (Y - X) = Y$$

and

$$X \cup Y = Y.$$

If X, Y, Z are sets, the equation

$$X \cap (Y \cup Z) = (X \cap Y) \cup (X \cap Z),$$

which expresses the distributivity of intersection over union, follows similarly from the distributivity of conjunction of propositions over their disjunction. The distributivity of union over intersection also follows by the same method.

In summary, we have the following theorem, which is analogous to Theorem 1.1.

Theorem 2.7 *Let X, Y, Z be sets. Then*

(a) $X \cup Y = Y \cup X$ *and* $X \cap Y = Y \cap X$, *and*

(b)
$$X \cap (Y \cup Z) = (X \cap Y) \cup (X \cap Z)$$

and

$$X \cup (Y \cap Z) = (X \cup Y) \cap (X \cup Z).$$

(c) *Moreover* $X \cap (Y - X) = \emptyset$, *and if* $X \subseteq Y$ *then* $X \cup (Y - X) = Y$.

(d) *Finally* $X \cup \emptyset = X$, *and if* $X \subseteq Y$ *then* $X \cap Y = X$.

The analogy with Theorem 1.1 becomes apparent if we replace the set $Y - X$ by a proposition s, then replace sets X, Y, Z by propositions p, q, r respectively, replace s by $\sim p$, replace equations, unions and intersections by equivalences, disjunctions and conjunctions respectively, replace $\emptyset$ by 0 and finally, if it is hypothesized that $X \subseteq Y$, replace q by 1. This analogy also reveals a duality similar to that already observed for propositions. Specifically, parts (a) and (b) of Theorem 2.7 remain valid if unions and intersections are interchanged. If $X \subseteq Y$ then parts (c) and (d) also remain valid after these interchanges provided that in addition Y and $\emptyset$ are interchanged in the right hand sides of the equations in (c) and in the left hand sides of the equations in (d).

We have already obtained equations analogous to 1.1 – 1.6 in Chapter 1. By combining these results with Theorem 2.7, using arguments analogous to the proofs of Theorems 1.2 and 1.4 and Lemma 1.3, we obtain the following theorem.

Theorem 2.8 *Let X, Y, Z be sets. Then*

(a) $X \cup (X \cap Y) = X \cap (X \cup Y) = X$, *and*
(b) $X \cup (Y \cup Z) = (X \cup Y) \cup Z$ *and* $X \cap (Y \cap Z) = (X \cap Y) \cap Z$.

Theorem 2.8 may also be viewed as a consequence of Theorems 1.2 and 1.4. Part (b) is of course the associative law for unions and intersections. In the light of this theorem, we may write $X \cup Y \cup Z$ instead of $X \cup (Y \cup Z)$, and $X \cap Y \cap Z$ instead of $X \cap (Y \cap Z)$.

We also have a theorem analogous to de Morgan's laws. In fact, it is an immediate consequence of them.

Theorem 2.9 *Let X and Y be subsets of a set Z. Then*

(a) $Z - (X \cup Y) = (Z - X) \cap (Z - Y)$ and
(b) $Z - (X \cap Y) = (Z - X) \cup (Z - Y)$.

For any set X whose elements are sets, we have defined the union $\cup X$ of X. Similarly we may define the **intersection** of X as the set of all $x \in \cup X$ such that $x \in Y$ for each set $Y \in X$. The intersection of X is denoted by $\cap X$. Since $\cap X \subseteq \cup X$ and $\cup \emptyset = \emptyset$, it follows that $\cap \emptyset = \emptyset$. If $x \in Y$ for each set $Y \in X$, then $x \in \cap X$ provided that $X \neq \emptyset$. Note also that

$$\cap \{X, Y\} = X \cap Y$$

for any sets X and Y.

Some more constructions of sets need to be mentioned. A set Y whose elements are sets is said to be **inductive** if

$$\emptyset \in Y$$

and

$$X \cup \{X\} \in Y$$

for each $X \in Y$. We hypothesize the existence of an inductive set Y. For example, Y contains $\emptyset \cup \{\emptyset\} = \{\emptyset\}$ and $\{\emptyset\} \cup \{\{\emptyset\}\} = \{\emptyset, \{\emptyset\}\}$. Intuitively, this hypothesis gives us an example of an infinite set. (Note however that we are not yet in a position to give a formal definition of an infinite set.) By replacing its elements by other sets, we may construct other "infinite" sets.

Finally, let X be a set whose elements are non-empty, mutually disjoint sets. Then we assert the existence of a set $Y \subseteq \cup X$ such that $Y \cap Z$ is a singleton for each set $Z \in X$. This assertion is commonly known as the **axiom of choice**, since it permits the selection of an object from each set in X. The axiom of choice plays an important rôle in advanced analysis.

We have now introduced all the collections of objects that we deem to be sets. It will be noted that our constructions of sets guarantee that the elements of any set are themselves sets. Indeed, this observation holds for $\emptyset$, which has no elements, and the elements of the other sets we have constructed are either sets by definition or elements of sets which have been constructed previously and which can therefore be assumed to enjoy the property that their elements are sets. Thus we have not defined the collection of all false propositions, for example, to be a set, since propositions are not sets.

2.2 Ordered Pairs, Relations and Functions

In this book, the concepts we need will be developed from the notion of a set, and results about these concepts will be proved by using the logical ideas presented in Chapter 1. Henceforth every object we contemplate, except in occasional informal discussions, will be a set of a type constructed in the previous section, even though we may occasionally call it a collection or an object for variety.

We now proceed to the development of more complicated concepts. The first of these is the notion of an ordered pair. If x and y are different objects (in other words, distinct sets), how can we get across the idea that x followed by y is different from y followed by x? If the former is to be denoted by (x,y) and the latter by (y,x), how can we define (x,y), in terms of sets, to ensure that $(x,y) \neq (y,x)$? The trick is to use x and y to construct a set which is altered if they are interchanged. One way of doing this is to define

$$(x,y) = \{\{x\}, \{x,y\}\}$$

for any objects x and y. We then call (x,y) an **ordered pair** with **components** x and y. We also say that x and y are the **first** and **second** components respectively.

In order to verify that the definition of an ordered pair (x,y) given above actually corresponds to our intuitive notion of x followed by y, we must show that ordered pairs (x,y) and (a,b) are equal if and only if $x = a$ and $y = b$. The sufficiency of this condition is obvious by substitution. Its necessity is the content of the following theorem.

Theorem 2.10 *If $(x,y) = (a,b)$ then $x = a$ and $y = b$.*

Proof. Since $(x,y) = (a,b)$, we have

$$\{\{x\}, \{x,y\}\} = \{\{a\}, \{a,b\}\} \tag{2.1}$$

by definition. We distinguish two cases.
Case I: Suppose $a = b$. Then from 2.1 we obtain

$$\begin{aligned}\{\{x\}, \{x,y\}\} &= \{\{a\}, \{a,a\}\} \\ &= \{\{a\}, \{a\}\} \\ &= \{\{a\}\}.\end{aligned}$$

Thus we must have $\{x,y\} = \{a\}$. Hence $x = a$ and $y = a = b$, as required.
Case II: Suppose $a \neq b$. From 2.1 we see that either $\{a,b\} = \{x\}$ or $\{a,b\} = \{x,y\}$. The former possibility is ruled out, however, as implying

the contradiction that $a = x = b$. Thus $\{a,b\} \neq \{x\}$, and so 2.1 implies not only that $\{x,y\} = \{a,b\}$ but also that $\{x\} = \{a\}$. The latter result gives $x = a$, as required, while from the former we infer that $b = x$ or $b = y$. But if $b = x$ we should have the contradiction that $b = a$. Therefore $y = b$. □

We can go further. For any objects x, y, z let us define
$$(x,y,z) = ((x,y),z).$$
Then (x,y,z) is called an **ordered triple**. Its **components** are x, y and z. As in the case of ordered pairs, we require a theorem to reconcile this definition with our intuitive notion of what ordered triples ought to be.

Theorem 2.11 *If $(x,y,z) = (a,b,c)$ then $x = a$, $y = b$ and $z = c$.*

Proof. Since $(x,y,z) = (a,b,c)$, we have
$$((x,y),z) = ((a,b),c).$$
Hence Theorem 2.10 gives $(x,y) = (a,b)$ and $z = c$. From the former equation we have $x = a$ and $y = b$, again by Theorem 2.10. □

Now suppose that X and Y are sets, and let $x \in X$ and $y \in Y$. Then $x \in X \cup Y$ and $y \in X \cup Y$, so that $\{x\} \subseteq X \cup Y$ and $\{x,y\} \subseteq X \cup Y$. Thus $\{x\} \in \mathcal{P}(X \cup Y)$ and $\{x,y\} \in \mathcal{P}(X \cup Y)$, and so
$$(x,y) = \{\{x\},\{x,y\}\} \subseteq \mathcal{P}(X \cup Y).$$
Hence
$$(x,y) \in \mathcal{P}(\mathcal{P}(X \cup Y)).$$
The set of all $(x,y) \in \mathcal{P}(\mathcal{P}(X \cup Y))$ such that $x \in X$ and $y \in Y$ is called the **Cartesian product** of X and Y, and denoted by $X \times Y$. For example, for any x and y we have $\{x\} \times \{x,y\} = \{(x,x),(x,y)\}$. Note also that
$$X \times \emptyset = \emptyset \times X = \emptyset$$
for any set X.

If X, Y, Z are sets, we write $X \times Y \times Z$ instead of $(X \times Y) \times Z$. The elements of $X \times Y \times Z$ are therefore ordered triples of the form $((x,y),z) = (x,y,z)$ where $x \in X$, $y \in Y$ and $z \in Z$.

Any subset of $X \times Y$ is called a **relation** from X to Y. It is simply a set of ordered pairs with first component in X and second component in Y. A subset of $X \times X$ is sometimes called a relation **on** X.

If R is a relation and x and y are objects, we occasionally write xRy instead of $(x,y) \in R$, and $x\not{R}y$ instead of $(x,y) \notin R$. As an example, let X be a set and let R be the relation on $\mathcal{P}(X)$ such that $(Y,Z) \in R$ if and only if Y is a subset of Z. (Of course Y and Z are subsets of X since R is a relation on $\mathcal{P}(X)$.) Then R is called the **inclusion** relation, and is denoted by $\subseteq$. We may then write $Y \subseteq Z$ instead of $(Y,Z) \in \subseteq$, in agreement with our earlier notation. Similarly there is a relation on $\mathcal{P}(X)$ whose elements are the ordered pairs $(Y,Z) \in \mathcal{P}(X) \times \mathcal{P}(X)$ such that Y is a proper subset of Z. This relation is called the **proper inclusion** relation, and is denoted by $\subset$, in agreement with our earlier notation. Again, the relation on $\mathcal{P}(X)$ whose elements are the ordered pairs $(Y,Z) \in \mathcal{P}(X) \times \mathcal{P}(X)$ such that Y and Z are equal is called the **equality** relation. In agreement with our earlier notation it is denoted by $=$.

Let R and S be relations and x, y, z objects. Then we may write $xRySz$ to indicate that $(x,y) \in R$ and $(y,z) \in S$. For example, if X, Y, Z are sets then we write $X \subseteq Y = Z$ if $X \subseteq Y$ and $Y = Z$. This notation may be extended to arbitrarily long chains of relations.

We proceed to describe various kinds of relations. Let R be a relation on a set X. We say that R is **reflexive** if $(x,x) \in R$ for each $x \in X$. Thus equality and inclusion are reflexive relations, but proper inclusion is not. The relation R is **symmetric** if $(y,x) \in R$ whenever $(x,y) \in R$. Thus equality is symmetric but inclusion and proper inclusion are not. The relation R is **antisymmetric** if $x = y$ whenever $(x,y) \in R$ and $(y,x) \in R$. Equality, inclusion and proper inclusion are all antisymmetric. (Proper inclusion is vacuously antisymmetric, since the hypothesis that X and Y are sets satisfying $X \subset Y \subset X$ is false.) The relation R is **transitive** if $(x,z) \in R$ whenever there exists $y \in X$ such that $(x,y) \in R$ and $(y,z) \in R$. Equality, inclusion and proper inclusion are all transitive.

A reflexive, antisymmetric, transitive relation on a set X is a **partial ordering** of X. Thus equality and inclusion are partial orderings, but proper inclusion is not reflexive.

A reflexive, symmetric, transitive relation on a set X is an **equivalence relation** on X. For example, equality is an equivalence relation, but inclusion and proper inclusion are not because they lack symmetry.

We pause to give an important characterization of equivalence relations. Let us first define a **partition** P of a non-empty set X to be a set of mutually disjoint, non-empty subsets of X whose union is X. The elements of P are sometimes called its **cells**. For instance, if $X = \{x,y\}$ where $x \neq y$, then $\{\{x\},\{y\}\}$ and $\{\{x,y\}\}$ are the only partitions of X. The cells of the former partition are $\{x\}$ and $\{y\}$, and $\{x,y\}$ is the only cell of the latter.

Equivalence relations give rise to partitions in a natural way, as we see in the following theorem.

Theorem 2.12 *Let R be an equivalence relation on a non-empty set X. Then there exists a partition P of X such that $(a,b) \in R$ if and only if a and b are elements of the same cell of P.*

Proof. For each $x \in X$ let $[x]$ be the set of all $y \in X$ such that $(x,y) \in R$. We shall show that the collection P of all sets of the form $[x]$, where $x \in X$, is a partition of X. Note that $x \in [x]$ since $(x,x) \in R$ by the reflexive property, and so $[x] \neq \emptyset$. An additional conclusion to be drawn from this observation is that $\cup P = X$. In order to show that P is a partition of X, it remains only to prove that the sets in P are mutually disjoint. For this purpose it suffices to show that $[x] = [y]$ whenever $x \in X$, $y \in X$ and $[x] \cap [y] \neq \emptyset$.

Suppose therefore that $x \in X$, $y \in X$ and there exists $z \in [x] \cap [y]$. Then $z \in [x]$ and $z \in [y]$. Since $z \in [y]$, we have $(y,z) \in R$. Similarly $(x,z) \in R$, and it follows that $(z,x) \in R$ since R is symmetric. Therefore $(y,x) \in R$ by the transitivity of R. In order to show that $[x] \subseteq [y]$, we choose $w \in [x]$. Thus $(x,w) \in R$, and so $(y,w) \in R$ by the transitive property again. Hence $w \in [y]$. We have now proved that $[x] \subseteq [y]$. By symmetry we also have $[y] \subseteq [x]$, and so $[x] = [y]$, as required.

Having checked that P is a partition of X, we must still show that $(a,b) \in R$ if and only if a and b are elements of the same cell of P. Choose $a \in [x]$, where $x \in X$. Then $(x,a) \in R$, and so $(a,x) \in R$ since R is symmetric. For each $b \in [x]$ we also have $(x,b) \in R$, and it follows that $(a,b) \in R$ by transitivity, as required. Conversely, suppose that $(a,b) \in R$ where $a \in [x]$. Since $(x,a) \in R$ we have $(x,b) \in R$ by transitivity. Hence $b \in [x]$, and the proof is complete. □

For each $x \in X$, the set $[x]$ defined in the proof above is called the **equivalence class** of X determined by x (under R). In other words, if R is an equivalence relation on a non-empty set X and $x \in X$, then $y \in [x]$ if and only if $(x,y) \in R$. We sometimes call x a **representative** of $[x]$. Theorem 2.12 shows that an equivalence relation on X is determined by the equivalence classes of X. In other words, once the equivalence classes are known, the equivalence relation is completely specified. Note also that each equivalence class is determined by any of its elements.

In order to complete our characterization of equivalence relations, we need a converse for Theorem 2.12. This converse is the content of the next theorem.

Theorem 2.13 *The cells of any partition of a non-empty set X are the equivalence classes of an equivalence relation on X.*

Proof. Let P be a partition of X, and let R be the relation on X such that $(x, y) \in R$ if and only if x and y belong to the same cell of P. It is immediate that R is an equivalence relation with the required property. □

These theorems are helpful, for they show that equivalence relations on a non-empty set X can be thought of as corresponding to partitions of X.

Let us now return to the general situation where we have sets X and Y and a relation F which is a subset of $X \times Y$. It may be that not every element of X is the first component of an ordered pair in F. However we may define the **domain**, $\mathcal{D}_F$, of F to be the set of all $x \in X$ for which there exists $y \in Y$ such that $(x, y) \in F$. Similarly the **range**, $\mathcal{R}_F$, of F is defined as the set of all $y \in Y$ for which there exists $x \in X$ such that $(x, y) \in F$. For instance, if $X = \{a, b, c\}$, $Y = \{d, e, f\}$ and $F = \{(a, f), (b, d)\}$, then $\mathcal{D}_F = \{a, b\}$ and $\mathcal{R}_F = \{d, f\}$.

Suppose now that $F \subseteq X \times Y$, that $\mathcal{D}_F = X$ and that $y = z$ whenever $(x, y) \in F$ and $(x, z) \in F$. Then F is a **function** from X into Y. For instance, in the example in the preceding paragraph, F is a function from $\{a, b\}$ into Y.

Thus a subset F of $X \times Y$ is a function from X into Y if and only if for each $x \in X$ there exists a unique $y \in Y$ such that $(x, y) \in F$. We call y the **image** of x under F, and write it as $F(x)$. (However if x is itself an ordered pair (a, b), then we usually write $F(a, b)$ instead of $F((a, b))$.) We may also say that F **maps** x to y, and that y **corresponds** to x under F. We may think of F as associating a unique object $y \in Y$ with each $x \in X$. Note that F is determined by its domain together with the image of each element of its domain. The situation may be pictured as in Figure 2.5, though it must be remembered that sets X and Y are not necessarily disjoint.

We sometimes write $F : X \to Y$ to indicate that F is a function from a set X into a set Y. We say that F is **constant** if $F(w) = F(x)$ for each $(w, x) \in X \times X$.

We have seen that if F is a function from X into Y then $\mathcal{D}_F = X$. If we also have $\mathcal{R}_F = Y$, then we may describe F as a **surjection** from X **onto** Y. This is the case if and only if for each $y \in Y$ there exists $x \in X$ such that $y = F(x)$.

We have also seen that if F is a function from X into Y then for each

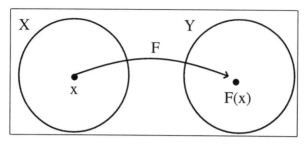

Fig. 2.5

$x \in X$ there exists a unique $y \in Y$ such that $(x, y) \in F$. In addition, for each $y \in \mathcal{R}_F$ there exists $x \in X$ such that $y = F(x)$. If this x is unique, then we say that F is an **injection** (or a **one-to-one** function) from X into Y. In this case we have $x = w$ whenever $F(x) = F(w)$. An injection from X into Y is called a **bijection** from X onto Y if it is also a surjection from X onto Y. Thus a function F from X into Y is a bijection from X onto Y if and only if for each $y \in Y$ there exists a unique $x \in X$ such that $y = F(x)$. A bijection from X onto Y is sometimes described as a 1 : 1 **correspondence** between X and Y.

As an example, let X be the set of all equivalence relations on a nonempty set Z, and let Y be the set of all partitions of Z. For each equivalence relation $R \in X$, let $F(R)$ be the partition of Z whose cells are the equivalence classes of Z under R. Since R is determined by the equivalence classes of Z, the function F is an injection. Theorem 2.13 shows that F is a surjection from X onto Y. Hence F is a bijection from X onto Y. In other words, we have established the existence of a 1 : 1 correspondence between the equivalence relations on Z and the partitions of Z.

Let X and Y be sets and F a relation which is a subset of $X \times Y$. Let F^{-1} be the subset of $Y \times X$ such that for any $x \in X$ and $y \in Y$ we have $(y, x) \in F^{-1}$ if and only if $(x, y) \in F$. Then the relation F^{-1} is the **inverse** of F. Note that $\mathcal{D}_{F^{-1}} = \mathcal{R}_F$. Clearly

$$(F^{-1})^{-1} = F.$$

Observe also that if F and F^{-1} are functions and $F(x) = y$ for some $x \in X$ and $y \in Y$, then $F^{-1}(y) = x$.

If F is a function, it is not necessarily true that F^{-1} is a function. However we have the following theorem.

Theorem 2.14 *If F is an injection then so is F^{-1}.*

Proof. We prove first that F^{-1} is a function. We must show that if $(x, y) \in F^{-1}$ and $(x, z) \in F^{-1}$ then $y = z$. Since $(x, y) \in F^{-1}$ we have $(y, x) \in F$. Similarly $(z, x) \in F$. That $y = z$ now follows from the fact that F is an injection. Hence F^{-1} is indeed a function.

In order to complete the proof, we must show that if $(x, z) \in F^{-1}$ and $(y, z) \in F^{-1}$ then $x = y$. Since $(x, z) \in F^{-1}$ we have $(z, x) \in F$. Similarly $(z, y) \in F$. It follows from the fact that F is a function that $x = y$, as required. □

If F is an injection from a set X into a set Y, then F^{-1} is an injection from $\mathcal{R}_F$ into X. A corollary of Theorem 2.14 is that if F is a bijection from a set X onto a set Y, then F^{-1} is a bijection from Y onto X. Indeed, $\mathcal{D}_{F^{-1}} = Y$ since $\mathcal{R}_F = Y$ (because F is a surjection from X onto Y), F^{-1} is an injection by Theorem 2.14, and $\mathcal{R}_{F^{-1}} = X$ since $\mathcal{D}_F = X$.

The situation where F is an injection from a set X into a set Y is depicted in Figure 2.6.

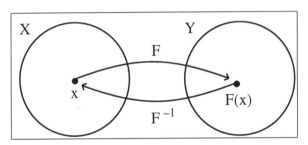

Fig. 2.6

Now let F be a function from a set X into a set Y, and let G be a function from Y into a set Z. Since $F(x) \in Y = \mathcal{D}_G$ for each $x \in X$, we may define $H(x) = G(F(x))$ for each $x \in X$. Therefore H is a function from X into Z. It is called the **composition** of G and F, and is denoted by $G \circ F$. Thus

$$(G \circ F)(x) = G(F(x))$$

for each $x \in X$. (See Figure 2.7.)

As an example, let F be an injection from a set X into a set Y. Suppose that $F(x) = y$ for some $x \in X$. Then

$$\begin{aligned}(F^{-1} \circ F)(x) &= F^{-1}(F(x)) \\ &= F^{-1}(y) \\ &= x.\end{aligned}$$

Similarly,

$$\begin{aligned}(F \circ F^{-1})(y) &= F(F^{-1}(y)) \\ &= F(x) \\ &= y.\end{aligned}$$

For any set X, we define I_X to be the function from X into X such that

$$I_X(x) = x$$

for each $x \in X$. Clearly I_X is a bijection from X onto X. It is called the **identity** function on X. If F is a function from a set X into a set Y, then for each $x \in X$ we have

$$\begin{aligned}(F \circ I_X)(x) &= F(I_X(x)) \\ &= F(x),\end{aligned}$$

so that

$$F \circ I_X = F.$$

Similarly

$$I_Y \circ F = F.$$

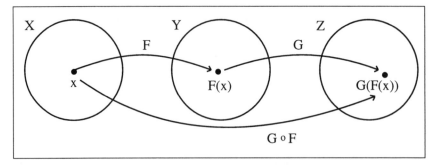

Fig. 2.7

The previous paragraph also shows that if F is a bijection from X onto Y then

$$F^{-1} \circ F = I_X$$

and

$$F \circ F^{-1} = I_Y.$$

The following theorem is the associative law for composition of functions.

Theorem 2.15 *Let F be a function from a set W into a set X, G a function from X into a set Y, and H a function from Y into a set Z. Then*

$$H \circ (G \circ F) = (H \circ G) \circ F.$$

Proof. Certainly $H \circ (G \circ F)$ is a function from W into Z, since $G \circ F$ is a function from W into Y and H is a function from Y into Z. Similarly $H \circ G$ is a function from X into Z, and since F is a function from W into X it follows that $(H \circ G) \circ F$ is also a function from W into Z. Moreover, for each $w \in W$ we have

$$\begin{aligned}(H \circ (G \circ F))(w) &= H((G \circ F)(w)) \\ &= H(G(F(w))) \\ &= (H \circ G)(F(w)) \\ &= ((H \circ G) \circ F)(w).\end{aligned}$$

Hence $H \circ (G \circ F) = (H \circ G) \circ F$. □

Thus we may write $H \circ G \circ F$ instead of $H \circ (G \circ F)$. This function is illustrated in Figure 2.8.

Theorem 2.16 *Let F be an injection from a set X into a set Y, and let G be an injection from Y into a set Z. Then $G \circ F$ is an injection from X into Z.*

Proof. Suppose there exist $x \in X$ and $y \in X$ such that

$$(G \circ F)(x) = (G \circ F)(y).$$

Then $G(F(x)) = G(F(y))$. Since G is an injection, it follows that $F(x) = F(y)$. Since F is an injection, we infer that $x = y$. Hence $G \circ F$ is an injection. □

40   Sets

Theorem 2.17 *Let F be a surjection from a set X onto a set Y, and let G be a surjection from Y onto a set Z. Then $G \circ F$ is a surjection from X onto Z.*

Proof. Choose $z \in Z$. Since G is a surjection from Y onto Z, there exists $y \in Y$ such that $G(y) = z$. Similarly there exists $x \in X$ such that $F(x) = y$. Therefore
$$\begin{aligned}(G \circ F)(x) &= G(F(x)) \\ &= G(y) \\ &= z.\end{aligned}$$
We conclude that $G \circ F$ is a surjection from X onto Z. □

By combining these two theorems we find that if F is a bijection from a set X onto a set Y and G is a bijection from Y onto a set Z, then $G \circ F$ is a bijection from X onto Z.

We round off this section with some more definitions that will be useful later. Let X be a set. Then a **unary operation** on X is a function from X into X, and a **binary operation** on X is a function from $X \times X$ into X. If F is a unary operation on X and $x \in X$, then we often write Fx instead of $F(x)$. Similarly if F is a binary operation on X and $(x, y) \in X \times X$, then we may write xFy instead of $F(x, y)$.

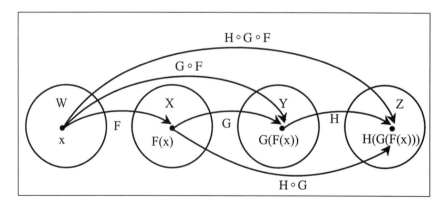

Fig. 2.8

Exercises 2

(1) (a) A set X is said to be **extraordinary** if $X \in X$ and **ordinary** otherwise. Show that the collection of all ordinary sets is not itself a set.
  (b) Let X be a set. Is the collection of all ordinary subsets of X a set?
(2) The **symmetric difference**, $X + Y$, of sets X and Y is defined as $(X \cup Y) - (X \cap Y)$.
  (a) For any sets X, Y, Z, prove that
   (i) $X + Y = Y + X$;
   (ii) $X + (Y + Z) = (X + Y) + Z$;
   (iii) $(X \cup Y) + (X \cup Z) \subseteq X \cup (Y + Z)$;
   (iv) $(X \cap Y) + (X \cap Z) = X \cap (Y + Z)$;
   (v) $(X + Y) \cap (X + Z) \subseteq X + (Y \cap Z)$;
   (vi) $(X + Y) \cap (X + Z) \subseteq X + (Y \cup Z) \subseteq (X + Y) \cup (X + Z)$.
  (b) Show that equality does not always hold in (iii), (v) and (vi).
(3) Let Z be the set of all partitions of a set U, and let $P \in Z$ and $Q \in Z$. We say that P is **finer** than Q if for each $X \in P$ there exists $Y \in Q$ for which $X \subseteq Y$. Let R be the relation on Z such that $(P, Q) \in R$ if and only if P and Q are partitions of U such that P is finer than Q. Show that R is a partial ordering of Z.
(4) A relation R on a set X is said to be **circular** if $(z, x) \in R$ whenever $(x, y) \in R$ and $(y, z) \in R$. Prove that R is an equivalence relation if and only if it is reflexive and circular.
(5) Does the commutative law hold for composition of functions? Either prove it or give a counterexample.
(6) Let f be a function from a set X into a set Y. For each $Z \subseteq X$ we define $f[Z]$ to be the set of all $f(x)$ for which $x \in Z$.
  (a) If $U \subseteq X$ and $W \subseteq X$, prove that:
   (i) $f[U \cup W] = f[U] \cup f[W]$;
   (ii) $f[U \cap W] \subseteq f[U] \cap f[W]$;
   (iii) $f[U] + f[W] \subseteq f[U + W]$.
  (b) Show that equality does not always hold in (ii) and (iii).

Chapter 3

Natural Numbers

3.1 Definition and Basic Properties

This chapter is devoted to using sets to construct the natural numbers — those employed to count objects in a given set. The discussion is based on the construction of an infinite set given in Chapter 2. In fact the natural numbers will be constructed from the empty set.

Recall that in Chapter 2 we hypothesized the existence of an inductive set: an inductive set Z contains $\emptyset$ and also contains $X \cup \{X\}$ for each set $X \in Z$. The next theorem asserts the existence of an inductive set with special properties.

Theorem 3.1 *There is a unique inductive set which is included in every inductive set.*

Proof. Choose an inductive set Z, and let Y be the set of all inductive sets included in Z. Define $\mathbb{N} = \cap Y$. We shall show that $\mathbb{N}$ is the required set.

First we prove that $\mathbb{N}$ is a subset of every inductive set. Let S be an inductive set, and choose $n \in \mathbb{N}$. Note that $\emptyset \in Z \cap S$, since both Z and S are inductive. Moreover, for each $X \in Z \cap S$ we have $X \in Z$ and $X \in S$, so that $X \cup \{X\} \in Z \cap S$. Therefore $Z \cap S$ is an inductive subset of Z, and so $Z \cap S \in Y$. Since $\mathbb{N} = \cap Y$ it follows that $\mathbb{N} \subseteq Z \cap S$, and as $n \in \mathbb{N}$ we conclude that $n \in Z \cap S$. Hence $n \in S$, and so $\mathbb{N} \subseteq S$.

Next we verify that $\mathbb{N}$ is inductive. Since $\emptyset$ is an element of every inductive set, we have $\emptyset \in \cap Y$. Thus $\emptyset \in \mathbb{N}$. Now choose $m \in \mathbb{N}$. Since $\mathbb{N} \subseteq S$ we have $m \in S$. But S is inductive, so that $m \cup \{m\} \in S$. As S is any inductive set, and the sets in Y are inductive, we infer that $m \cup \{m\} \in \cap Y$. Hence $m \cup \{m\} \in \mathbb{N}$, and we deduce that $\mathbb{N}$ is inductive.

Finally, suppose that M is an inductive set which is included in every inductive set. Then $M \subseteq \mathbb{N}$ and $\mathbb{N} \subseteq M$, so that $M = \mathbb{N}$. Therefore $\mathbb{N}$ is unique. □

Henceforth we shall reserve the letter $\mathbb{N}$ for the unique inductive set which is included in every inductive set.

The elements of $\mathbb{N}$ are called **natural numbers**. The natural number $\emptyset$ is denoted by 1, $\emptyset \cup \{\emptyset\} = \{\emptyset\} = \{1\}$ by 2 and $\{\emptyset\} \cup \{\{\emptyset\}\} = \{\emptyset, \{\emptyset\}\} = \{1, 2\}$ by 3. We now develop the elementary properties of the natural numbers.

First, for each $n \in \mathbb{N}$ we have $n \cup \{n\} \in \mathbb{N}$. Let us define

$$S(n) = n \cup \{n\}$$

for each $n \in \mathbb{N}$. Then S is a unary operation on $\mathbb{N}$. The number $S(n)$ is called the **successor** of n, and will also be denoted by n^+. For example, $1^+ = 1 \cup \{1\} = \{1\} = 2$. Similarly $2^+ = 2 \cup \{2\} = \{1\} \cup \{2\} = \{1, 2\} = 3$. The function S will be called the **successor function**.

Frequently in mathematics a class of objects may be defined by giving a particular member A of the class and then specifying operations which may be performed on existing members of the class to give new ones. A property shared by all members of the class may then be proved by showing first that it holds for A and then that it is preserved by the specified operations. When applying the latter step to a particular operation, we assume that the desired property holds for the objects on which the operation is to be performed, and then we prove the property for any new members of the class that are created by the operation. Our construction of sets in Chapter 2 gives an example: we introduced $\emptyset$ first (in Chapter 1) and then gave various methods of constructing new sets from old. Similarly in Chapter 1 we introduced a false proposition which asserted the membership in $\emptyset$ of a given object, and then we constructed new propositions from old ones by the negation and disjunction operations and the use of the existential quantifier.

The inductive set $\mathbb{N}$ furnishes yet another example of this idea. We have $1 \in \mathbb{N}$, and for each $n \in \mathbb{N}$ we have specified how to construct a new member n^+ of $\mathbb{N}$. In order to prove a theorem about each $n \in \mathbb{N}$, it suffices to establish the theorem for 1 and then show that it holds for n^+ whenever it holds for a particular $n \in \mathbb{N}$. This technique is a very powerful one for proving theorems about natural numbers, and is referred to as **induction** (on n). The assumption that the theorem holds for a particular n is often called the **inductive hypothesis**.

Definition and Basic Properties   45

We now state the concept of induction more formally as a theorem.

Theorem 3.2 *Let M be a subset of $\mathbb{N}$ such that $1 \in M$. Suppose also that $n^+ \in M$ whenever $n \in M$. Then $M = \mathbb{N}$.*

Proof. The hypotheses show that M is an inductive set. Since $\mathbb{N}$ is included in every inductive set, we deduce that $\mathbb{N} \subseteq M$. But $M \subseteq \mathbb{N}$, and so $M = \mathbb{N}$. □

The technique of induction can be viewed as a special case of Theorem 3.2: we take M to be the set of all natural numbers for which a given predicate p holds, and we try to prove that p holds for all natural numbers (in other words, that $M = \mathbb{N}$) by verifying that M satisfies the hypotheses of the theorem.

As an application of induction, we prove in the next theorem that every natural number $n \neq 1$ is the successor of a natural number.

Theorem 3.3 *S is a surjection from $\mathbb{N}$ onto $\mathbb{N} - \{1\}$.*

Proof. Let $M = \{1\} \cup \mathcal{R}_S$. Thus $M \subseteq \mathbb{N}$ and $1 \in M$. Next, choose $m \in M$. Since $m^+ = S(m)$, we have $m^+ \in \mathcal{R}_S$. Hence $m^+ \in M$. It follows by induction that $\mathbb{N} = M = \{1\} \cup \mathcal{R}_S$. Since $S(n) \neq \emptyset$ for each $n \in \mathbb{N}$, we find that $1 \notin \mathcal{R}_S$. Therefore $\mathcal{R}_S = \mathbb{N} - \{1\}$, as required. □

So far so good, but we would really like to prove that S is an injection. In other words, we would like to show that if m and n are distinct natural numbers then $m^+ \neq n^+$. Then we would know that each natural number except 1 is the successor of a unique natural number. In order to make progress on this question, we need another idea.

A set Z (whose elements are sets) is **transitive** if a set X is contained in Z whenever there exists a set Y such that $X \in Y$ and $Y \in Z$. Thus if Z is transitive and $Y \in Z$, then we have $X \in Z$ whenever $X \in Y$, so that $Y \subseteq Z$. Conversely if $Y \subseteq Z$ whenever $Y \in Z$ then Z is transitive: if $X \in Y$ and $Y \in Z$ then $Y \subseteq Z$, so that $X \in Z$. In other words, a set Z is transitive if and only if each element is a subset of Z.

Transitive sets may seem difficult to visualize. However the following lemma shows that each natural number is an example.

Lemma 3.4 *Every natural number is transitive.*

Proof. Let M be the set of transitive natural numbers.

Since $\emptyset$ has no elements, it is vacuously true that each element is a subset of $\emptyset$. Hence $\emptyset$ is transitive, and so $1 \in M$.

Choose $n \in M$. To complete the proof by induction, we must show that n^+ is transitive. Choose $m \in n^+ = n \cup \{n\}$ and $l \in m$. In order to show that $m \subseteq n^+$, we must prove that $l \in n^+$. There are two possibilities: either $m \in n$ or $m \in \{n\}$. In the former case we have $l \in n$ since n is transitive. In the latter case $m = n$, and so $l \in n$ once again. In both cases $l \in n^+$. We conclude that $m \subseteq n^+$, so that n^+ is transitive, as required. □

We can now obtain the result we want.

Theorem 3.5 *S is an injection.*

Proof. We need to rule out the possibility that there could exist distinct natural numbers m and n for which $S(m) = S(n)$. Suppose therefore that $m \cup \{m\} = n \cup \{n\}$ and $m \neq n$. Since $m \in m \cup \{m\} = n \cup \{n\}$ but $m \neq n$, we must have $m \in n$. But n is transitive by Lemma 3.4, and so $m \subseteq n$. A similar argument shows that $n \subseteq m$. Thus we have the contradiction that $m = n$, and the proof is complete. □

Thus each natural number $n \neq 1$ is the successor of a unique natural number, which is called the **predecessor** of n.

We complete our study of the properties of the successor function with the following theorem.

Theorem 3.6 *For each $n \in \mathbb{N}$ we have $n^+ \neq n$.*

Proof. Since $1^+ \neq \emptyset = 1$, the theorem holds if $n = 1$. Assume as an inductive hypothesis that n is a natural number for which the theorem holds. Then $n^+ \neq n$. Therefore $n^{++} \neq n^+$ by Theorem 3.5. In other words, the theorem holds for n^+, and the proof is complete. □

Thus no natural number is the successor of itself.

Theorems 3.3, 3.5 and 3.6 enable us to list the natural numbers in order so that 1 is the first one and each is followed by its successor. We have already seen that the first three, in order, are denoted by 1, 2 and 3 respectively. The next six, in order, are denoted by 4, 5, 6, 7, 8, 9 respectively. A notation for subsequent members of $\mathbb{N}$ will be developed in Chapter 4.

Our next theorem, which we will call the recursion theorem, requires some explanation. Essentially it introduces a process of definition of a function by induction which is analogous to proof by induction. Specifically, let F be a unary operation on a set X, and let $x \in X$. (Thus F is a function from X into X.) It may be that we wish to select a succession of elements of X, beginning with x and then applying F repeatedly. This selection is accomplished by constructing a function H from $\mathbb{N}$ into X so that $H(1) = x$ and if H maps a natural number n to an object $y \in X$ then H maps n^+ to $F(y)$. In other words, $H(S(n)) = F(H(n))$ for each $n \in \mathbb{N}$. (See Figure 3.1.)

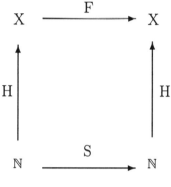

Fig. 3.1

Intuitively, the essence of this idea is that H relates the natural numbers to our selections of elements of X in such a way that successive natural numbers are related to successive selections. The content of the recursion theorem is that there exists a unique function H with the required property.

Theorem 3.7 *Let F be a unary operation on a set X and let $x \in X$. Then there exists a unique function H from $\mathbb{N}$ into X such that $H(1) = x$ and $H \circ S = F \circ H$.*

Proof. We define $H = \cap T$ where T is the set of all $W \subseteq \mathbb{N} \times X$ satisfying the conditions that $(1, x) \in W$ and $(n^+, F(y)) \in W$ whenever $(n, y) \in W$. One of our objectives is to prove that H is a function from $\mathbb{N}$ into X. However we shall first prove that $T \neq \emptyset$. For this purpose it suffices to show that $\mathbb{N} \times X \in T$.

Certainly $\mathbb{N} \times X \subseteq \mathbb{N} \times X$ and $(1, x) \in \mathbb{N} \times X$. Choose $(n, y) \in \mathbb{N} \times X$. Then $n \in \mathbb{N}$ and $y \in X$. Hence $n^+ \in \mathbb{N}$ and $F(y) \in X$, and it follows that $(n^+, F(y)) \in \mathbb{N} \times X$. Therefore $\mathbb{N} \times X \in T$ by the definition of T. Hence $T \neq \emptyset$.

We show next that $H \in T$. Since H is an intersection of subsets of $\mathbb{N} \times X$, we clearly have $H \subseteq \mathbb{N} \times X$. Moreover $(1, x) \in H$ because $T \neq \emptyset$ and $(1, x) \in W$ for each $W \in T$. Finally, if $(n, y) \in H = \cap T$ then $(n, y) \in W$ for each $W \in T$. Therefore $(n^+, F(y)) \in W$ for each $W \in T$, and so

$(n^+, F(y)) \in \cap T = H$. We conclude that $H \in T$.

In what follows we repeatedly use the fact that $H \subseteq W$ for each $W \in T$, since $H = \cap T$.

In order to show that H is a function from $\mathbb{N}$ into X, we need to demonstrate that for each $n \in \mathbb{N}$ there exists a unique $a \in X$ for which $(n, a) \in H$. The proof will be by induction on n.

We already know that $(1, x) \in H$. Suppose that x is not the only element of X with this property. Then there exists $b \in X - \{x\}$ such that $(1, b) \in H$. Define $H_b = H - \{(1, b)\}$. We shall prove that $H_b \in T$. Certainly $H_b \subseteq H \subseteq \mathbb{N} \times X$, and $(1, x) \in H_b$ since $x \neq b$. Suppose that $(n, y) \in H_b$. We must show that $(n^+, F(y)) \in H_b$. We have $(n, y) \in H$, and as $H \in T$ it follows that $(n^+, F(y)) \in H$. Moreover $(n^+, F(y)) \neq (1, b)$ since $n^+ \neq 1$, and so $(n^+, F(y)) \in H - \{(1, b)\} = H_b$, as required. Therefore $H_b \in T$. Thus we reach the contradiction that $H \subseteq H_b$, since $H \subseteq W$ for each $W \in T$. The conclusion is that x must be unique.

We now take as our inductive hypothesis that for some $n \in \mathbb{N}$ there exists a unique $a \in X$ such that $(n, a) \in H$. Since $H \in T$, it follows that $(n^+, F(a)) \in H$. We must show that $F(a)$ is the only element of X with this property. If not, there must exist $c \in X - \{F(a)\}$ such that $(n^+, c) \in H$. Define $H_c = H - \{(n^+, c)\}$.

We now verify that $H_c \in T$. Certainly $H_c \subseteq \mathbb{N} \times X$, and $(1, x) \in H_c$ since $n^+ \neq 1$. Choose $(m, z) \in H_c$. We show that $(m^+, F(z)) \in H_c$. Since $(m, z) \in H_c \subseteq H$ and $H \in T$, it follows that $(m^+, F(z)) \in H$. We must rule out the possibility that $(m^+, F(z)) = (n^+, c)$. Should this equation hold, then $m^+ = n^+$ and $F(z) = c \neq F(a)$. The former result would show that $m = n$ (by Theorem 3.5), so that $(n, z) \in H$, but from the latter result we should have $z \neq a$. These conclusions would contradict the uniqueness of a in the inductive hypothesis. Therefore $(m^+, F(z)) \neq (n^+, c)$, and so $(m^+, F(z)) \in H - \{(n^+, c)\} = H_c$.

The proof that $H_c \in T$ is now complete. We therefore have the contradiction that $H \subseteq H_c$. Hence $F(a)$ is indeed the only element $v \in X$ for which $(n^+, v) \in H$.

We have now proved by induction that H is a function from $\mathbb{N}$ into X. In addition, $H(1) = x$ since $(1, x) \in H$.

Next we show that $H \circ S = F \circ H$. For any $(n, y) \in H$ we have $H(n) = y$

and $(n^+, F(y)) \in H$. Thus $H(n^+) = F(y)$, so that

$$\begin{aligned}(H \circ S)(n) &= H(S(n)) \\ &= H(n^+) \\ &= F(y) \\ &= F(H(n)) \\ &= (F \circ H)(n)\end{aligned}$$

for each $n \in \mathbb{N}$. Hence $H \circ S = F \circ H$.

It remains only to establish the uniqueness of H. Let G be a function from $\mathbb{N}$ into X such that $G(1) = x$ and $G \circ S = F \circ G$. We shall prove that $G = H$ by using induction to show that $G(n) = H(n)$ for each $n \in \mathbb{N}$. Certainly $G(1) = x = H(1)$. Suppose as an inductive hypothesis that $G(n) = H(n)$ for some $n \in \mathbb{N}$. Then

$$\begin{aligned}G(n^+) &= G(S(n)) \\ &= (G \circ S)(n) \\ &= (F \circ G)(n) \\ &= F(G(n)) \\ &= F(H(n)) \\ &= (F \circ H)(n) \\ &= (H \circ S)(n) \\ &= H(S(n)) \\ &= H(n^+).\end{aligned}$$

Thus $G(n) = H(n)$ for each $n \in \mathbb{N}$ by induction, and so $G = H$. The uniqueness of H has now been established. □

The following application of the recursion theorem will enable us to introduce some binary operations on $\mathbb{N}$. (Recall that a binary operation on $\mathbb{N}$ is a function from $\mathbb{N} \times \mathbb{N}$ into $\mathbb{N}$.)

Theorem 3.8 *Let F be a unary operation on $\mathbb{N}$, and for each $m \in \mathbb{N}$ let F_m be a unary operation on $\mathbb{N}$. Then there is a unique binary operation H on $\mathbb{N}$ such that*

$$H(m, 1) = F(m)$$

for each $m \in \mathbb{N}$ *and*

$$H(m, n^+) = F_m(H(m, n))$$

for each $(m, n) \in \mathbb{N} \times \mathbb{N}$.

Remark: Given $m \in \mathbb{N}$, the function H specifies how m is to be combined with other natural numbers so as to define a binary operation on $\mathbb{N}$. In the next sections we shall give three examples, yielding binary operations which we shall call addition, multiplication and exponentiation respectively.

Proof. We prove the existence of H first. Choose $m \in \mathbb{N}$. By the recursion theorem (applied to the unary operation F_m on $\mathbb{N}$ and taking $x = F(m)$), we find that there exists a unique unary operation H_m on $\mathbb{N}$ such that $H_m(1) = F(m)$ and $H_m \circ S = F_m \circ H_m$. For each $(m, n) \in \mathbb{N} \times \mathbb{N}$ define $H(m, n) = H_m(n)$. Then

$$H(m, 1) = H_m(1) = F(m)$$

and

$$\begin{aligned} H(m, n^+) &= H_m(n^+) \\ &= H_m(S(n)) \\ &= (H_m \circ S)(n) \\ &= (F_m \circ H_m)(n) \\ &= F_m(H_m(n)) \\ &= F_m(H(m, n)), \end{aligned}$$

as required.

Now let G be any binary operation on $\mathbb{N}$ such that $G(m, 1) = F(m)$ for each $m \in \mathbb{N}$ and $G(m, n^+) = F_m(G(m, n))$ for each $(m, n) \in \mathbb{N} \times \mathbb{N}$. In order to establish the uniqueness of H, we must show that $H = G$. It suffices to prove that $H(m, n) = G(m, n)$ for each $(m, n) \in \mathbb{N} \times \mathbb{N}$. Choose $m \in \mathbb{N}$. We prove that $H(m, n) = G(m, n)$ for each $n \in \mathbb{N}$ by induction on n. Certainly $H(m, 1) = F(m) = G(m, 1)$. Suppose as an inductive hypothesis that $H(m, n) = G(m, n)$ for some $n \in \mathbb{N}$. Then

$$\begin{aligned} H(m, n^+) &= F_m(H(m, n)) \\ &= F_m(G(m, n)) \\ &= G(m, n^+), \end{aligned}$$

as required. Hence $H = G$. □

3.2 Addition

Recall that a binary operation on a set X is a function from $X \times X$ into X. Thus a binary operation on $\mathbb{N}$ maps each ordered pair of natural numbers to a unique natural number. We shall study several such operations, the first of which will be called addition.

Theorem 3.9   *There is a unique binary operation H on $\mathbb{N}$ such that*

$$H(m, 1) = m^+$$

for each $m \in \mathbb{N}$ and

$$H(m, n^+) = (H(m, n))^+$$

for each $(m, n) \in \mathbb{N} \times \mathbb{N}$.

Remark: The motivation for the latter equation is that the addition of m and n^+ should give the successor of the natural number obtained by the addition of m and n.

Proof.   Apply Theorem 3.8 with $F_m = F = S$ for each $m \in \mathbb{N}$.   □

The binary operation H introduced in Theorem 3.9 is called **addition**, and $H(m, n)$ is called the **sum** of the natural numbers m and n. We generally write $m + n$ instead of $H(m, n)$. In this notation, the equations in Theorem 3.9 show that

$$m^+ = m + 1$$

for each $m \in \mathbb{N}$, and that

$$\begin{aligned} m + (n + 1) &= m + n^+ \\ &= (m + n)^+ \\ &= (m + n) + 1 \end{aligned} \qquad (3.1)$$

for each $m \in \mathbb{N}$ and $n \in \mathbb{N}$.

We proceed to the basic properties of this operation. The first such property is associativity.

Theorem 3.10   *If $(k, m, n) \in \mathbb{N} \times \mathbb{N} \times \mathbb{N}$ then*

$$k + (m + n) = (k + m) + n.$$

Proof. We already know from 3.1 that the result holds for $n = 1$. Assume therefore that k and m are fixed and, as an inductive hypothesis, that the theorem then holds for some $n \in \mathbb{N}$. We obtain

$$\begin{aligned} k + (m + (n+1)) &= k + ((m+n) + 1) & \text{(by 3.1)} \\ &= (k + (m+n)) + 1 & \text{(by 3.1)} \\ &= ((k+m) + n) + 1 \\ &= (k+m) + (n+1), \end{aligned}$$

again by 3.1. The theorem follows by induction. □

We usually write $k + m + n$ instead of $k + (m + n)$. As an example we have

$$\begin{aligned} 2 + 2 &= 2 + (1 + 1) \\ &= (2 + 1) + 1 \\ &= 3 + 1 \\ &= 4. \end{aligned}$$

Commutativity also holds. We prove it first in a special case.

Lemma 3.11 *If $n \in \mathbb{N}$ then $n + 1 = 1 + n$.*

Proof. Clearly the lemma holds if $n = 1$. Assume as an inductive hypothesis that $n + 1 = 1 + n$ for some $n \in \mathbb{N}$. Then

$$\begin{aligned} (n + 1) + 1 &= (1 + n) + 1 \\ &= 1 + (n + 1) \end{aligned}$$

by associativity. Thus the lemma follows by induction. □

Theorem 3.12 *If $(m, n) \in \mathbb{N} \times \mathbb{N}$ then $m + n = n + m$.*

Proof. For each $m \in \mathbb{N}$ we use induction on n. Lemma 3.11 shows that the result holds for $n = 1$. Suppose as an inductive hypothesis that $m + n = n + m$ for some $n \in \mathbb{N}$. Then

$$\begin{aligned} m + (n + 1) &= (m + n) + 1 \\ &= (n + m) + 1 \\ &= n + (m + 1) \\ &= n + (1 + m) \\ &= (n + 1) + m. \end{aligned}$$

The theorem follows by induction. □

The next theorem is known as the **cancellation law**.

Theorem 3.13 *If $(k, m, n) \in \mathbb{N} \times \mathbb{N} \times \mathbb{N}$ and $k+n = m+n$, then $k = m$.*

Proof. We use induction on n. For $n = 1$ the theorem is immediate from the fact that the successor function S is an injection. Assume as an inductive hypothesis that the theorem holds for some $n \in \mathbb{N}$, and suppose that
$$k + n + 1 = m + n + 1.$$
Again since S is an injection we have $k + n = m + n$, whence $k = m$ by the inductive hypothesis. The theorem follows by induction. □

Theorem 3.14 *If $(m, n) \in \mathbb{N} \times \mathbb{N}$ then $n \neq m + n$.*

Proof. The result holds for $n = 1$ since $1 \notin \mathcal{R}_S$. Fix $m \in \mathbb{N}$ and suppose that $n \neq m + n$ for some $n \in \mathbb{N}$. Then $n + 1 \neq m + n + 1$ since S is an injection, and the theorem follows by induction. □

3.3 Multiplication

Our next operation will be called multiplication.

Theorem 3.15 *There is a unique binary operation H on $\mathbb{N}$ such that*
$$H(m, 1) = m$$
for each $m \in \mathbb{N}$ and
$$H(m, n^+) = H(m, n) + m$$
for each $(m, n) \in \mathbb{N} \times \mathbb{N}$.

Remark: The motivation for the latter equation is that the multiplication of m and $n^+ = n + 1$ should give the sum of m and the result of the multiplication of m and n.

Proof. Apply Theorem 3.8 with $F = I_\mathbb{N}$ and $F_m(n) = n + m$ for each $(m, n) \in \mathbb{N} \times \mathbb{N}$. □

This binary operation is called **multiplication**, and $H(m, n)$ is called the **product** of the natural numbers m and n. We generally write $m \cdot n$, or simply mn, instead of $H(m, n)$. Thus we have
$$m \cdot 1 = m$$

for each $m \in \mathbb{N}$, and

$$m(n+1) = mn + m \qquad (3.2)$$

for each $(m, n) \in \mathbb{N} \times \mathbb{N}$. The operation of multiplication has priority over addition.

As an example, we have

$$\begin{aligned}
2 \cdot 2 &= 2(1+1) \\
&= 2 \cdot 1 + 2 \qquad \text{(by 3.2)}\\
&= 2 + 2 \\
&= 4.
\end{aligned}$$

The following theorem shows that multiplication is distributive over addition.

Theorem 3.16 If $(k, m, n) \in \mathbb{N} \times \mathbb{N} \times \mathbb{N}$ then

$$(k+m)n = kn + mn.$$

Proof. We fix k and m and use induction on n. The theorem holds for $n = 1$, since

$$(k+m) \cdot 1 = k + m = k \cdot 1 + m \cdot 1.$$

Suppose as an inductive hypothesis that it holds for some $n \in \mathbb{N}$. Then

$$\begin{aligned}
(k+m)(n+1) &= (k+m)n + k + m \\
&= (kn + mn) + (k + m) \\
&= ((kn + mn) + k) + m \\
&= (kn + (mn + k)) + m \\
&= (kn + (k + mn)) + m \\
&= ((kn + k) + mn) + m \\
&= (kn + k) + (mn + m) \\
&= k(n+1) + m(n+1).
\end{aligned}$$

The theorem follows by induction. □

Again, we establish commutativity by proving a special case first.

Lemma 3.17 For each $n \in \mathbb{N}$ we have $1 \cdot n = n \cdot 1$.

Proof. The lemma is clear if $n = 1$. Suppose that $1 \cdot n = n \cdot 1$ for some $n \in \mathbb{N}$. Then

$$\begin{aligned} 1 \cdot (n+1) &= 1 \cdot n + 1 \\ &= n \cdot 1 + 1 \\ &= n + 1 \\ &= (n+1) \cdot 1, \end{aligned}$$

as required. The lemma follows by induction. □

Theorem 3.18 *If $(m, n) \in \mathbb{N} \times \mathbb{N}$ then $mn = nm$.*

Proof. By Lemma 3.17 the theorem holds if $n = 1$. Fix m and suppose that $mn = nm$ for some $n \in \mathbb{N}$. Then

$$\begin{aligned} m(n+1) &= mn + m \qquad \text{(by 3.2)} \\ &= mn + m \cdot 1 \\ &= nm + 1 \cdot m \\ &= (n+1)m, \end{aligned}$$

by Theorem 3.16. The proof by induction is complete. □

Multiplication is also associative.

Theorem 3.19 *If $(k, m, n) \in \mathbb{N} \times \mathbb{N} \times \mathbb{N}$, then $k(mn) = (km)n$.*

Proof. We fix $k \in \mathbb{N}$ and $m \in \mathbb{N}$ and use induction on n. The theorem holds for $n = 1$, since

$$k(m \cdot 1) = km = (km) \cdot 1.$$

Now suppose as an inductive hypothesis that $k(mn) = (km)n$ for some $n \in \mathbb{N}$. Then

$$\begin{aligned} k(m(n+1)) &= k(mn + m) \qquad \text{(by 3.2)} \\ &= k(mn) + km \\ &= (km)n + km \\ &= (km)(n+1), \end{aligned}$$

by 3.2, as required. □

3.4 Exponentiation

Theorem 3.20 *There is a unique binary operation H on $\mathbb{N}$ such that*

$$H(m, 1) = m$$

for each $m \in \mathbb{N}$ and

$$H(m, n^+) = mH(m, n)$$

for each $(m, n) \in \mathbb{N} \times \mathbb{N}$.

Proof. Apply Theorem 3.8 with $F = I_\mathbb{N}$ and $F_m(n) = mn$ for each $(m, n) \in \mathbb{N} \times \mathbb{N}$. $\square$

This binary operation is called **exponentiation**, and $H(m, n)$ is called the nth **power** of m. We generally write m^n instead of $H(m, n)$. Thus we have

$$m^1 = m$$

for each $m \in \mathbb{N}$, and

$$m^{n+1} = m \cdot m^n \tag{3.3}$$

for each $(m, n) \in \mathbb{N} \times \mathbb{N}$. Thus exponentiation is related to multiplication in the same way as the latter is related to addition.

For example,

$$2^2 = 2^{1+1} = 2 \cdot 2^1 = 2 \cdot 2 = 4.$$

Note that $1^1 = 1$ and if $1^n = 1$ for some $n \in \mathbb{N}$ then

$$1^{n+1} = 1 \cdot 1^n = 1 \cdot 1 = 1.$$

Hence $1^n = 1$ for all $n \in \mathbb{N}$ by induction.

We proceed to the elementary properties of exponentiation. The first of these is distributivity over multiplication. Its proof is analogous to that of Theorem 3.16, and is therefore omitted.

Theorem 3.21 *For each $(k, m, n) \in \mathbb{N} \times \mathbb{N} \times \mathbb{N}$ we have*

$$(km)^n = k^n m^n.$$

Theorem 3.22 *For each $(k, m, n) \in \mathbb{N} \times \mathbb{N} \times \mathbb{N}$ we have*
$$k^m k^n = k^{m+n}.$$

Proof. The theorem holds for $n = 1$, since
$$\begin{aligned} k^{m+1} &= k^m k \quad &\text{(by 3.3)} \\ &= k^m k^1. \end{aligned}$$

Now fix $k \in \mathbb{N}$ and $m \in \mathbb{N}$ and suppose that $k^m k^n = k^{m+n}$ for some $n \in \mathbb{N}$. Then
$$\begin{aligned} k^{m+n+1} &= k^{m+n} k \quad &\text{(by 3.3)} \\ &= k^m k^n k \\ &= k^m k^{n+1}, \end{aligned}$$

by 3.3, and the proof by induction is complete. □

Theorem 3.23 *For each $(k, m, n) \in \mathbb{N} \times \mathbb{N} \times \mathbb{N}$ we have*
$$(k^m)^n = k^{mn}.$$

Proof. The theorem clearly holds for $n = 1$. Fix $k \in \mathbb{N}$ and $m \in \mathbb{N}$ and suppose that $(k^m)^n = k^{mn}$ for some $n \in \mathbb{N}$. Then
$$\begin{aligned} (k^m)^{n+1} &= (k^m)^n k^m \quad &\text{(by 3.3)} \\ &= k^{mn} k^m \\ &= k^{mn+m} \\ &= k^{m(n+1)}, \end{aligned}$$

and the proof is complete by induction. □

3.5 Order

Let $m \in \mathbb{N}$ and $n \in \mathbb{N}$. We say that m is **less** than n, or that n is **greater** than m, if there exists $k \in \mathbb{N}$ such that $m + k = n$. (Hence k is unique, by the cancellation law.) In this case we write $m < n$ or $n > m$. The set of all $(m, n) \in \mathbb{N} \times \mathbb{N}$ such that $m < n$ thus defines a relation (called an **inequality**) on $\mathbb{N}$. In agreement with our earlier notation, we therefore write $m \not< n$ if m is not less than n. For example, $n \not< n$ for each $n \in \mathbb{N}$, since $n + k \neq n$ for each $(k, n) \in \mathbb{N} \times \mathbb{N}$ by Theorem 3.14. (Thus if $m < n$ or $m > n$ then $m \neq n$.) Hence this relation is certainly not reflexive. However it is transitive, as we see in the following result.

Theorem 3.24 *If $k < m$ and $m < n$, then $k < n$.*

Proof. There exist $x \in \mathbb{N}$ such that $k + x = m$ and $y \in \mathbb{N}$ such that $m + y = n$. Hence $k + x + y = n$, so that $k < n$. □

In keeping with our notation concerning transitive relations, we therefore write $k < m < n$ if $k < m$ and $m < n$. Note that if $m < n$ then $n \not< m$, for otherwise we obtain the contradiction that $m < n < m$. Thus each of the possibilities $m < n$, $m = n$ and $m > n$ rules out the others. However we also have the following theorem, known as the **trichotomy property**.

Theorem 3.25 *If $(m, n) \in \mathbb{N} \times \mathbb{N}$ then either $m < n$, $m = n$ or $m > n$.*

Proof. We fix m and use induction on n. Suppose first that $n = 1$. If $m = 1$ then $m = n$. If $m > 1$ then, by Theorem 3.3, there exists $k \in \mathbb{N}$ such that $m = k + 1 = k + n$. In this case $m > n$.

We now assume as an inductive hypothesis that for some $n \in \mathbb{N}$ either $m < n$, $m = n$ or $m > n$.
Case I: Suppose $m < n$. Then there exists $k \in \mathbb{N}$ such that $m + k = n$. Hence $m + k + 1 = n + 1$, and so $m < n + 1$.
Case II: Suppose $m = n$. Then $m + 1 = n + 1$, and again $m < n + 1$.
Case III: Finally suppose that $m > n$. This time there exists $k \in \mathbb{N}$ such that $n + k = m$. If $k = 1$ we have $m = n + 1$. Otherwise by Theorem 3.3 there exists $l \in \mathbb{N}$ such that $k = l + 1$. Then $n + 1 + l = m$ and we conclude that $n + 1 < m$.

The proof is complete in all cases by induction. □

We note that $n > 1$ for each $n \in \mathbb{N} - \{1\}$, since we may write $n = m + 1$ for some $m \in \mathbb{N}$ by Theorem 3.3.

For each $m \in \mathbb{N}$ and $n \in \mathbb{N}$, we write $m \leq n$ or $n \geq m$ if either $m < n$ or $m = n$. For example, the observation above shows that $n \geq 1$ for each $n \in \mathbb{N}$. Clearly we also have $n \leq n$. If $k < m$ and $m \leq n$, then $k < n$ (by Theorem 3.24 in the case where $m < n$). Similarly if $k \leq m$ and $m < n$ then again $k < n$. The transitivity of this relation follows, for if $k \leq m$, $m \leq n$ and $k \neq n$, then either $k < m$ or $m < n$ by Theorem 2.2(c), and we deduce that $k < n$ from the results above. Note also that if $m \not\leq n$, then $n < m$ by Theorem 3.25. Moreover if $m \leq n \leq m$ then we must have $m = n$ to avert the contradiction that $m < m$. This relation is therefore a partial ordering.

Theorem 3.26 *Let $(k, m, n) \in \mathbb{N} \times \mathbb{N} \times \mathbb{N}$. Then $m < n$ if and only if $m + k < n + k$.*

In other words, an equal amount can be added to both sides of an inequality.

Proof. If $m < n$ then $n = m + l$ for some $l \in \mathbb{N}$. Therefore $m + k + l = n + k$, and so $m + k < n + k$.

Conversely, suppose $m + k < n + k$. Reversal of the argument above then shows that $m < n$. □

Some immediate consequences of this theorem deserve to be noted. First, $m \leq n$ if and only if $m + k \leq n + k$ for each $k \in \mathbb{N}$. Next, if $k < l$ and $m < n$, then

$$k + m < k + n < l + n.$$

It follows that if $k < l$ and $m \leq n$ then $k + m < l + n$. Similarly this inequality also holds if $k \leq l$ and $m < n$. Thus if $k \leq l$ and $m \leq n$ then $k + m \leq l + n$. Indeed, if $k = l$ and $m = n$ then $k + m = l + n$; otherwise $k + m < l + n$. Also, suppose that $m < n$. Then there exists $k \in \mathbb{N}$ such that $m + k = n$. Since $1 \leq k$, it follows that $m + 1 \leq n$. Thus if $m + 1 > n$, then $m \geq n$.

Let us move on to multiplication.

Theorem 3.27 *Let $(k, m, n) \in \mathbb{N} \times \mathbb{N} \times \mathbb{N}$. Then $m < n$ if and only if $km < kn$.*

Proof. If $m < n$ then $n = m + l$ for some $l \in \mathbb{N}$. Therefore $kn = km + kl$, and so $km < kn$. Conversely, suppose that $km < kn$. If $m = n$ then $km = kn$, and if $n < m$ then $kn < km$. Both these results contradict our hypothesis. Hence $m < n$. □

An immediate corollary is the cancellation law for multiplication: if $km = kn$ then $m = n$, since the other possibilities lead to immediate contradictions by Theorem 3.27. It also follows that $m \leq n$ if and only if $km \leq kn$. If $k < l$ and $m < n$, then $km < kn < ln$. Thus if $k < l$ and $m \leq n$, then $km < ln$. Similarly $km < ln$ if $k \leq l$ and $m < n$. Therefore if $k \leq l$ and $m \leq n$ then $km \leq ln$.

Finally we consider exponentiation.

Theorem 3.28 *If $(k, m, n) \in \mathbb{N} \times \mathbb{N} \times \mathbb{N}$, then $k^n < m^n$ if and only if $k < m$.*

Proof. Suppose $k < m$. We proceed by induction on n. There is nothing to prove if $n = 1$. Assume as an inductive hypothesis that $k^n < m^n$ for

some $n \in \mathbb{N}$. Then
$$k^{n+1} = k \cdot k^n < m \cdot m^n = m^{n+1},$$
and so $k^n < m^n$ in general, by induction.

Conversely, if $k^n < m^n$ then $k < m$, as the other possibilities contradict the hypothesis. □

Theorem 3.29 *Let $(k, m, n) \in \mathbb{N} \times \mathbb{N} \times \mathbb{N}$. Then $k^m < k^n$ if and only if $k > 1$ and $m < n$.*

Proof. Suppose $k > 1$ and $m < n$. Then $n = m + l$ for some $l \in \mathbb{N}$. Moreover $1 = 1^l < k^l$ by Theorem 3.28, and so
$$k^m = k^m \cdot 1 < k^m k^l = k^{m+l} = k^n.$$

Conversely if $k^m < k^n$ then $k > 1$, since $1^m = 1^n = 1$. Furthermore $m < n$, as the other possibilities contradict our hypothesis. □

For example, if $k > 1$ and $n > 1$ then $k^n > k$.

Let $M \subseteq \mathbb{N}$. A number $m \in M$ is called a **least** element of M if $m \leq n$ for each $n \in M$. Clearly m is unique, for if $m \leq n$ and $k \leq n$ for each $n \in M$, where $m \in M$ and $k \in M$, then $m \leq k$ and $k \leq m$, so that $m = k$.

Theorem 3.30 *Let $M \subseteq \mathbb{N}$ and $M \neq \emptyset$. Then M has a least element.*

Proof. Let K be the set of all $k \in \mathbb{N}$ such that $k \leq m$ for each $m \in M$. Thus $1 \in K$. Moreover $K \neq \mathbb{N}$, for if we choose $m \in M$ then $m + 1 \notin K$ since $m < m+1$. Therefore there exists $k \in K$ such that $k+1 \notin K$, because otherwise we should have $K = \mathbb{N}$ by induction.

Recall that $k \leq m$ for each $m \in M$. In order to demonstrate that k is the least element of M, it therefore remains only to show that $k \in M$. Otherwise $k < m$ for each $m \in M$. Hence $k + 1 \leq m$ for each $m \in M$, and so we have the contradiction that $k + 1 \in K$. Therefore $k \in M$, and k is the required least element of M. □

We are now able to give a stronger version of induction.

Theorem 3.31 *Let $M \subseteq \mathbb{N}$. Suppose that M contains every natural number n such that $m \in M$ for each natural number $m < n$. Then $M = \mathbb{N}$.*

Proof. Suppose $M \neq \mathbb{N}$. Then $\mathbb{N} - M \neq \emptyset$. By Theorem 3.30, $\mathbb{N} - M$ has a least element n. For each $m < n$ it follows that $m \notin \mathbb{N} - M$, so that $m \in M$. Therefore by hypothesis we obtain the contradiction that $n \in M$. We conclude that $M = \mathbb{N}$. □

Thus in order to prove a theorem about natural numbers, it suffices to show that the theorem holds for a natural number n whenever it holds for each natural number less than n. The assumption that the theorem holds for *each* natural number less than n is more powerful than the inductive hypothesis in our previous version of induction. Accordingly this version of induction is more potent.

3.6 Applications

In this section we use the recursion theorem to generalize some of the concepts we encountered earlier.

Let X be a set and $+$ a binary operation on X satisfying the associative law. This operation will be referred to as **addition**. The ordered pair $(X, +)$ is then called a **semigroup**. For example, $(\mathcal{P}(X), \cup)$, $(\mathcal{P}(X), \cap)$, $(\mathbb{N}, +)$ and $(\mathbb{N}, \cdot)$ are semigroups. The semigroup is **commutative** if the operation $+$ is commutative. The examples above are all commutative semigroups.

If $(X, +)$ is a semigroup and $(x, y, z) \in X \times X \times X$, then associativity shows that $x + (y + z) = (x + y) + z$. Accordingly we usually write $x + y + z$ instead of $x + (y + z)$. Similarly if w, x, y, z are elements of X then, by associativity,

$$w + (x + y + z) = (w + x) + (y + z) = (w + x + y) + z,$$

and so we may write $w + x + y + z$ instead of $(w + x) + (y + z)$. This idea will be extended later to a generalized associative law.

At the moment, however, we have a different generalization in mind. Note first that

$$w + x + y = (w + x) + y$$

and

$$w + x + y + z = (w + x + y) + z.$$

We aim to continue this process indefinitely. The recursion theorem provides the means.

Let G be a function from $\mathbb{N}$ into X. Thus G maps each natural number to a member of X. Let F be the unary operation on $\mathbb{N} \times X$ such that

$$F(n, x) = (n + 1, x + G(n + 1))$$

for each $(n, x) \in \mathbb{N} \times X$. According to the recursion theorem, for each $m \in \mathbb{N}$ there exists a unique function H_m from $\mathbb{N}$ into $\mathbb{N} \times X$ such that $H_m(1)$ is the element $(m, G(m))$ of $\mathbb{N} \times X$ and $H_m \circ S = F \circ H_m$. Thus

$$H_m(n+1) = F(H_m(n))$$

for each $n \in \mathbb{N}$. For instance we have

$$H_m(1) = (m, G(m)),$$
$$H_m(2) = F(H_m(1))$$
$$= F(m, G(m))$$
$$= (m+1, G(m) + G(m+1))$$

and

$$H_m(3) = F(H_m(2))$$
$$= F(m+1, G(m) + G(m+1))$$
$$= (m+2, G(m) + G(m+1) + G(m+2)).$$

Next, for each $n \in \mathbb{N}$ define $P(n, n) = G(n)$, and for each natural number $m < n$ define $P(m, n)$ to be the second component of $H_m(l+1)$, where l is the unique natural number such that $m + l = n$. For example:

$$P(1, 3) = G(1) + G(2) + G(3);$$
$$P(2, 3) = G(2) + G(3);$$
$$P(3, 3) = G(3).$$

The following lemma shows that the function P so defined encapsulates the property we desire.

Lemma 3.32 *For each $(m, n) \in \mathbb{N} \times \mathbb{N}$ such that $m \leq n$ we have*

$$P(m, n+1) = P(m, n) + G(n+1).$$

Proof. If $m = n$ then $m + l = n + 1$ for $l = 1$, and $P(m, n+1)$ is the second component of $H_m(2)$, which is

$$G(m) + G(m+1) = G(n) + G(n+1)$$
$$= P(n, n) + G(n+1),$$

as required.

On the other hand, suppose $m < n$. Then $m + l = n$ for some natural number l, so that $m+l+1 = n+1$, and $P(m, n+1)$ is the second component of $H_m(l+2)$.

We show by induction on l that the first component of $H_m(l+1)$ is $m + l = n$. Indeed, the first component of $H_m(2)$ is $m + 1$, and if

$$H_m(l+1) = (m+l, P(m,n))$$

for some $l \in \mathbb{N}$ then the first component of

$$\begin{aligned} H_m(l+2) &= F(H_m(l+1)) \\ &= F(m+l, P(m,n)) \end{aligned}$$

is $m + l + 1$, as required. Since the second component of

$$H_m(l+2) = F(n, P(m,n))$$

is $P(m,n) + G(n+1)$, the lemma holds in this case also. □

We now define

$$\sum_{k=m}^{n} G(k) = P(m,n)$$

for each $(m,n) \in \mathbb{N} \times \mathbb{N}$ such that $m \leq n$. Thus

$$\sum_{k=n}^{n} G(k) = G(n)$$

and, from Lemma 3.32,

$$\sum_{k=m}^{n+1} G(k) = \sum_{k=m}^{n} G(k) + G(n+1).$$

Observe that k is a dummy variable in the notation above: we have

$$\sum_{k=m}^{n} G(k) = \sum_{j=m}^{n} G(j)$$

for any letters j and k.

In the semigroups $(\mathcal{P}(X), \cup)$, $(\mathcal{P}(X), \cap)$ and $(\mathbb{N}, \cdot)$, the notation $\sum_{k=m}^{n} G(k)$ is replaced by $\cup_{k=m}^{n} G(k)$, $\cap_{k=m}^{n} G(k)$ and $\prod_{k=m}^{n} G(k)$ respectively. Each $G(k)$ such that $m \leq k \leq n$ is called a **summand** or **term** of $\sum_{k=m}^{n} G(k)$ and a **factor** of $\prod_{k=m}^{n} G(k)$. We say that $\sum_{k=m}^{n} G(k)$ is formed by the **addition** of its summands, and is their **sum** or **total**.

Similarly $\prod_{k=m}^{n} G(k)$ is formed by the **multiplication** of its factors, and it is their **product**.

If $n = m + 1$, then from Lemma 3.32 we have

$$\sum_{k=m}^{m+1} G(k) = \sum_{k=m}^{m} G(k) + G(m+1)$$
$$= G(m) + G(m+1).$$

Thus $\sum_{k=m}^{n} G(k)$ is a generalization of our earlier notion of addition. As special cases we also have

$$\bigcup_{k=m}^{m+1} G(k) = G(m) \cup G(m+1), \tag{3.4}$$

$$\bigcap_{k=m}^{m+1} G(k) = G(m) \cap G(m+1)$$

and

$$\prod_{k=m}^{m+1} G(k) = G(m)G(m+1).$$

For $\prod_{k=1}^{n} k$, where $n \in \mathbb{N}$, we often write $n!$. This number is called the **factorial** of n. For example, $3! = 3 \cdot 2 \cdot 1 = 6$.

The following theorem gives a useful interpretation of $\bigcup_{k=m}^{n} G(k)$ and $\bigcap_{k=m}^{n} G(k)$. For each $(m, n) \in \mathbb{N} \times \mathbb{N}$ such that $m \leq n$, we let $[m, n]$ denote the set of all $k \in \mathbb{N}$ for which $m \leq k \leq n$. We may refer to $[m, n]$ as an **interval** of natural numbers.

Theorem 3.33 *Let $(m, n) \in \mathbb{N} \times \mathbb{N}$, where $m \leq n$, and let X be a set. Let G be a function from $[m, n]$ into $\mathcal{P}(X)$. Then*

(a) $\bigcup_{k=m}^{n} G(k)$ *is the set of all $x \in X$ for which there exists $k \in [m, n]$ such that $x \in G(k)$, and*
(b) $\bigcap_{k=m}^{n} G(k)$ *is the set of all $x \in X$ such that $x \in G(k)$ for each k in $[m, n]$.*

Proof. (a) The result is clear if $m = n$. Suppose therefore that $m < n$. Then $m + l = n$ for some natural number l. We argue by induction on l. We see from 3.4 that the result holds if $l = 1$. We therefore suppose as an inductive hypothesis that

$$\bigcup_{k=m}^{m+l} G(k)$$

is the set of all $x \in X$ for which there exists $k \in [m, m+l]$ satisfying $x \in G(k)$.

Now we consider

$$\bigcup_{k=m}^{m+l+1} G(k) = \bigcup_{k=m}^{m+l} G(k) \cup G(m+l+1).$$

Choose

$$x \in \bigcup_{k=m}^{m+l+1} G(k).$$

Either $x \in \cup_{k=m}^{m+l} G(k)$ or $x \in G(m+l+1)$. In the former case the inductive hypothesis yields the existence of a natural number $k \in [m, m+l]$ such that $x \in G(k)$. Thus in both cases there exists $k \in [m, m+l+1]$ such that $x \in G(k)$.

Conversely, suppose such a k exists for some $x \in X$. If $k = m+l+1$, then $x \in G(m+l+1)$; otherwise $m \le k \le m+l$, so that $x \in \cup_{k=m}^{m+l} G(k)$ by the inductive hypothesis. We conclude in both cases that

$$x \in \bigcup_{k=m}^{m+l+1} G(k),$$

as required.

The proof of (b) is similar. □

Recall that $2 = \{1\}$ and $3 = \{1, 2\}$. It is easily seen by induction that any $n \in \mathbb{N}$ is the set of all natural numbers less than n. Indeed, $1 = \emptyset$ by definition. Suppose as an inductive hypothesis that n is the set of all $m \in \mathbb{N}$ such that $m < n$. Since $n + 1 = n \cup \{n\}$, it follows from the inductive hypothesis that a natural number m is an element of $n + 1$ if and only if $m \le n$, as required. In terms of our notation for an interval of natural numbers, we also have $n + 1 = [1, n]$.

Note also that if $(m, n) \in \mathbb{N} \times \mathbb{N}$ and $m < n$ (so that $m \in n$), then $m \subset n$. Indeed, $m \subseteq n$ by the transitivity of n, and since $m \in n$ but $m \notin m$ it follows that $m \subset n$, as we claimed. Moreover, if $m \le n$ then $m \subseteq n$.

A set X is said to be **finite** if for some natural number n there exists a bijection from n onto X. Otherwise we describe X as **infinite**.

We shall show that the natural number n in the definition above, if it exists, is unique. First, let us define sets X and Y to be **equinumerous** (with each other) if there is a bijection from X onto Y. Thus a finite set is equinumerous with some natural number. Any set is equinumerous with itself, as the identity function supplies the required bijection. If X and Y are equinumerous, then so are Y and X, for if F is a bijection from X onto Y then F^{-1} is the required bijection from Y onto X. We have also seen that if F is a bijection from X onto Y and G is a bijection from Y onto another set Z, then $G \circ F$ is a bijection from X onto Z. It follows that if X is equinumerous with Y, and Y with Z, then X is equinumerous with Z.

It is possible for a given set to be equinumerous with a proper subset. For instance, the successor function shows that $\mathbb{N}$ is equinumerous with $\mathbb{N} - \{1\}$. In fact an inductive argument shows that $\mathbb{N}$ is equinumerous with $\mathbb{N} - n$ for any $n \in \mathbb{N}$. However we do have the following theorem.

Theorem 3.34 *No natural number is equinumerous with any of its proper subsets.*

Proof. We use induction. As 1 has no proper subset, we may assume as an inductive hypothesis that some natural number n is not equinumerous with any of its proper subsets. Suppose however that $n + 1$ does have a proper subset M equinumerous with it. Thus there is a bijection F from $n + 1$ onto M. Hence $F(k) \leq n$ for each $k \leq n$.

Suppose first that $F(k) < n$ for each $k < n$. We shall show that n is equinumerous with $M - \{F(n)\}$. Let G be the function from n into $M - \{F(n)\}$ such that $G(k) = F(k)$ for each $k < n$. Clearly G is an injection since F is. For each $m \in M - \{F(n)\}$ there exists $k < n$ such that $m = F(k) = G(k)$ since F is a surjection from $n + 1$ onto M. Therefore G is a surjection onto $M - \{F(n)\}$, and hence a bijection. Thus n is indeed equinumerous with $M - \{F(n)\}$.

Next we prove that $M - \{F(n)\} \subset n$. Choose $m \in M$. If $m \neq F(n)$, then $m = F(k)$ for some $k < n$. Hence $m < n$ by hypothesis, so that $m \in n$. Therefore $M - \{F(n)\} \subseteq n$. If $F(n) \in n$, then in fact $M - \{F(n)\} \subset n$. Otherwise $F(n) = n$, so that $n \in M$. In this case there exists $k < n$ such that $k \notin M$, since $M \subset n + 1$, and again it follows that $M - \{F(n)\} \subset n$. Thus we have the contradiction that n is equinumerous with one of its proper subsets.

We conclude that there exists $m < n$ for which $F(m) = n$. Thus

$F(k) < n$ for each $k \neq m$.

We show next that n is equinumerous with $M - \{n\}$. Define $H(m) = F(n)$ and $H(k) = F(k)$ for each $k < n$ such that $k \neq m$. Then H is a function from n into $M - \{n\}$. It is an injection since F is. To see that H is a surjection onto $M - \{n\}$, choose $k \in M - \{n\}$. We must find $j < n$ such that $H(j) = k$. There exists $l \leq n$ such that $F(l) = k$ since F is a surjection, and $l \neq m$ since $F(m) = n$. If $l = n$ then $H(m) = F(n) = F(l) = k$, as required. If $l < n$ then $H(l) = F(l) = k$ since $l \neq m$. We conclude that H is indeed a bijection onto $M - \{n\}$.

Hence n is equinumerous with $M - \{n\}$. But $M \subset n+1$, and so $M - \{n\} \subseteq n$. Moreover $F(m) = n$, so that $n \in M$. Hence there exists $p < n$ such that $p \notin M$, since $M \subset n+1$, and we deduce that $M - \{n\} \subset n$. Again we have the contradiction that n is equinumerous with one of its proper subsets.

We conclude that $n + 1$ cannot be equinumerous with any of its proper subsets. The proof is now complete by induction. □

Corollary 3.35 *Any finite set has a unique natural number equinumerous with it.*

Proof. By definition any finite set X is equinumerous with some natural number n. Suppose there exists $m \neq n$ such that X is also equinumerous with m. Without loss of generality we may assume that $m < n$. Then $m \subset n$. By Theorem 3.34, m is therefore not equinumerous with n, in contradiction to the fact that X is equinumerous with both. □

Let X be a finite set. By Corollary 3.35 there exists a unique natural number n equinumerous with X. Suppose now that $X \neq \emptyset$. Then $n > 1$, since there is no bijection from $1 = \emptyset$ onto X. Therefore $n = m + 1$ for some unique $m \in \mathbb{N}$. We call m the **number** of elements of X, or the **cardinality** of X, and we say that X has m elements. The cardinality of X is denoted by $|X|$.

An application of the results of this section enables us to list the elements of a non-empty finite set X. There is a bijection F from $n+1$ onto X, where $n = |X|$. For each $k \leq n$ we define $x_k = F(k)$. Since F is a surjection onto X, each element $x \in X$ is of the form x_k for some $k \leq n$. Thus $x \in \{x_k\}$. It follows from Theorem 3.33(a) that

$$X = \bigcup_{k=1}^{n} \{x_k\}.$$

For example, if $n = 1$ then

$$X = \bigcup_{k=1}^{1} \{x_k\} = \{x_1\},$$

and if $n = 2$ then

$$X = \bigcup_{k=1}^{2} \{x_k\} = \{x_1\} \cup \{x_2\} = \{x_1, x_2\}.$$

In general we write $\cup_{k=1}^{n} \{x_k\}$ as $\{x_1, x_2, \cdots, x_n\}$, and we may list the elements of X as $x_1, x_2, \cdots, x_n$.

Now let X be a set equinumerous with $\mathbb{N}$. There exists a bijection F from $\mathbb{N}$ onto X. Again let us denote $F(n)$ by x_n for each $n \in \mathbb{N}$. We may then write X as $\{x_1, x_2, \cdots\}$.

For any non-empty finite set X, we may evaluate $|X|$ by counting the elements of X. This procedure amounts to an application of the formula

$$\sum_{k=1}^{n} 1 = n,$$

which can be proved by induction for each natural number n. Indeed, $\sum_{k=1}^{1} 1 = 1$, and if $\sum_{k=1}^{n} 1 = n$ for some $n \in \mathbb{N}$ then

$$\sum_{k=1}^{n+1} 1 = \sum_{k=1}^{n} 1 + 1 = n + 1,$$

as required.

If X is a non-empty finite set and Y is a non-empty subset of X, then Y is also finite and $|Y| \leq |X|$. Indeed, X is equinumerous with some natural number n and Y is equinumerous with a non-empty subset of n and therefore with some natural number $m \leq n$. It follows that if X is infinite and $X \subseteq Y$ then Y is infinite.

Now suppose that X and Y are non-empty disjoint finite sets. Then $X \cup Y$ is finite and

$$|X \cup Y| = |X| + |Y|,$$

for if $X = \{x_1, x_2, \cdots, x_m\}$ and $Y = \{y_1, y_2, \cdots, y_n\}$ then $X \cup Y$ is the finite set

$$\{x_1, x_2, \cdots, x_m, y_1, y_2, \cdots, y_n\}.$$

We now turn to an extension of the concepts of an ordered pair and an ordered triple. In fact, we shall define an **ordered n-tuple** $(x_1, x_2, \cdots, x_n)$ for each $n \in \mathbb{N}$, where $x_1, x_2, \cdots, x_n$ are objects which are called the **components** of the ordered n-tuple. The definition is by induction. We begin by defining $(x_1) = x_1$. Next, we suppose that $(x_1, x_2, \cdots, x_n)$ has already been defined for some natural number n, and we proceed to define $(x_1, x_2, \cdots, x_{n+1})$. Recall that (x_1, x_2) has already been defined in Section 2.2, and so we may suppose that $n > 1$. We then define

$$(x_1, x_2, \cdots, x_{n+1}) = ((x_1, x_2, \cdots, x_n), x_{n+1}),$$

in agreement with our earlier definition of (x_1, x_2, x_3). We now have the following theorem.

Theorem 3.36 *For each $n \in \mathbb{N}$, we have*

$$(x_1, x_2, \cdots, x_n) = (y_1, y_2, \cdots, y_n) \tag{3.5}$$

if and only if $x_k = y_k$ for each $k \leq n$.

Proof. Clearly 3.5 holds if $x_k = y_k$ for each $k \leq n$. Moreover the converse of this statement holds for $n = 1$. We therefore suppose as an inductive hypothesis that this converse holds for a natural number n, and we try to prove it for $n + 1$. Thus we assume that

$$(x_1, x_2, \cdots, x_{n+1}) = (y_1, y_2, \cdots, y_{n+1}),$$

and we prove that $x_k = y_k$ for each $k \leq n+1$. By Theorem 2.10 we may assume that $n > 1$. Then by definition we have

$$((x_1, x_2, \cdots, x_n), x_{n+1}) = ((y_1, y_2, \cdots, y_n), y_{n+1}),$$

and so we deduce from Theorem 2.10 that

$$(x_1, x_2, \cdots, x_n) = (y_1, y_2, \cdots, y_n)$$

and $x_{n+1} = y_{n+1}$. That $x_k = y_k$ for each $k \leq n$ now follows from the inductive hypothesis. The proof is complete by induction. □

If $X_1, X_2, \cdots, X_n$ are sets for some $n > 1$, then

$$X_1 \times X_2 \times \cdots \times X_n$$

denotes the set of all ordered n-tuples of the form $(x_1, x_2, \cdots, x_n)$ where $x_k \in X_k$ for each $k \leq n$. This set is called the **Cartesian product** of

$X_1, X_2, \cdots, X_n$. It was defined earlier in the special case where $n = 2$, and we complete the definition inductively, setting

$$X_1 \times X_2 \times \cdots \times X_{n+1} = (X_1 \times X_2 \times \cdots \times X_n) \times X_{n+1}.$$

Let F be a function mapping $X_1 \times X_2 \times \cdots \times X_n$ into a set X. If $x_k \in X_k$ for each $k \leq n$, then we write $F(x_1, x_2, \cdots, x_n)$ instead of $F((x_1, x_2, \cdots, x_n))$.

Exercises 3

(1) In Theorem 3.7, set $X = \mathbb{N}$, $x = 2$ and $F(n) = n + 2$ for each $n \in \mathbb{N}$. Find a formula for $H(n)$ for each $n \in \mathbb{N}$.

(2) Prove the existence of H in Theorem 3.7 as follows. First, for any $n \in \mathbb{N}$ define a function K from a set M into a set X to be n-**admissible** if (i) $\{1, n\} \subseteq M \subseteq \mathbb{N}$, (ii) $K(1) = x$ and (iii) if $m^+ \in M$ for some $m \in \mathbb{N}$ then $m \in M$ and $K(m^+) = F(K(m))$.

  (a) Prove that an n^+-admissible function is n-admissible.
  (b) Use induction to prove that there exists an n-admissible function for each $n \in \mathbb{N}$.
  (c) Show that if G and K are n-admissible then $G(n) = K(n)$.
  (d) Use (a) – (c) to deduce the existence of H.

(3) Evaluate $2 + 3$, $2 \cdot 3$, 2^3 and 3^2 from the definitions.

(4) Prove that if $n > 1$ then $n^m > m$ for each m.

(5) Prove that

$$(m + n)^2 = m^2 + 2mn + n^2$$

and

$$(m + n)^3 = m^3 + 3m^2n + 3mn^2 + n^3$$

for any natural numbers m and n.

(6) Prove that $(1 + m)^n \geq 1 + mn$ for any natural numbers m and n.

(7) A natural number is **even** if it is of the form $2n$ for some $n \in \mathbb{N}$, and **odd** if it is of the form $2n + 1$ for some $n \in \mathbb{N}$.

  (a) Prove that each natural number is either odd or even but not both.
  (b) Determine the circumstances under which (i) $m + n$ and (ii) mn are odd, where $m \in \mathbb{N}$ and $n \in \mathbb{N}$.

(8) (a) If X is a finite set, prove that $X \cup \{x\}$ is finite for any object x.

(b) Show that any subset of a finite set is finite.
(9) For any natural number $n \geq 4$ prove that $2^n < n!$.
(10) Prove that $2^n > n$ for any natural number n.

Chapter 4

Integers

4.1 Definition

If m and n are natural numbers and $m < n$, then there exists $l \in \mathbb{N}$ such that $m + l = n$. This number l is sometimes written as $n - m$. Thus $m + (n - m) = n$. However there is no natural number l such that $1 + l = 1$, since $1 \notin \mathcal{R}_S$. Moreover, if $m > 1$ then we may write $m = k + 1$ for some $k \in \mathbb{N}$. Therefore there cannot be a natural number l such that $m + l = 1$, for otherwise we should have the contradiction that $1 + k + l = 1$.

These observations lead us to devise numbers other than the natural ones. The idea is to define a new kind of number which is a set of ordered pairs of natural numbers. We begin by defining an equivalence relation on $\mathbb{N} \times \mathbb{N}$. If $(k, l) \in \mathbb{N} \times \mathbb{N}$ and $(m, n) \in \mathbb{N} \times \mathbb{N}$, we take $(k, l) \sim (m, n)$ to mean that $k + n = l + m$. For example, $(3, 1) \sim (4, 2)$ and $(4, 2) \sim (5, 3)$, since $3 + 2 = 1 + 4 = 5$ and $4 + 3 = 2 + 5 = 7$. If $(k, l) \sim (m, n)$ and $m > n$, then it follows from Theorem 3.26 that $l < k$. Moreover

$$l + (k - l) + n = k + n$$
$$= l + m$$
$$= l + n + (m - n),$$

so that $k - l = m - n$ by cancellation. In general, if $(k, m, n) \in \mathbb{N} \times \mathbb{N} \times \mathbb{N}$ then $(m, n) \sim (m + k, n + k)$ since $m + n + k = n + m + k$.

Theorem 4.1 *The relation $\sim$ is an equivalence relation on $\mathbb{N} \times \mathbb{N}$.*

Proof. It is immediate from the commutativity of addition that $(m, n) \sim (m, n)$ for each $(m, n) \in \mathbb{N} \times \mathbb{N}$. In other words, reflexivity holds. Symmetry follows similarly from the symmetry of equality and the commutativity of addition. Suppose therefore that $(i,j) \sim (k,l)$ and $(k,l) \sim (m,n)$. Then $i+l = j+k$ and $k+n = l+m$. Hence

$$i+l+k+n = j+k+l+m,$$

and by the cancellation law it follows that $i+n = j+m$. In other words $(i,j) \sim (m,n)$ and we have proved transitivity. □

We let $\mathbb{Z}$ be the set of equivalence classes under $\sim$. The elements of $\mathbb{Z}$ are called **integers**. For example, $[(3,1)] = \{(3,1),(4,2),\cdots\}$. We write $0 = [(1,1)]$ and $1 = [(2,1)]$. Thus we now have two meanings for "1". They will be reconciled later, but for the time being the reader should ensure that he understands which meaning is intended whenever the symbol "1" is used. Note that 0 and 1 are distinct integers, for $(1,1) \in 0$ but $(1,1) \notin 1$ since $1+1 = 2$ and $2+1 = 3 \neq 2$. Furthermore we have $(m,n) \in 0$ if and only if $(m,n) \sim (1,1)$. By the cancellation law it follows that $(m,n) \in 0$ if and only if $m = n$, since $(m,n) \sim (1,1)$ if and only if $m+1 = n+1$.

4.2 Addition

Before introducing the idea of addition for integers, we prove the following lemma. It is needed in order to guarantee that the concept of addition about to be introduced is in fact well defined.

Lemma 4.2 *Let g, h, i, j, k, l, m, n be natural numbers. If $(g,h) \sim (k,l)$ and $(i,j) \sim (m,n)$, then*

$$(g+i, h+j) \sim (k+m, l+n).$$

Remark: If $k > l$ and $m > n$, then

$$k+m = l + (k-l) + n + (m-n),$$

so that

$$(k-l) + (m-n) = (k+m) - (l+n).$$

This equation is the reason for our interest in the lemma.

Proof. We are given that $g+l = h+k$ and $i+n = j+m$. Therefore

$$\begin{aligned} g+i+l+n &= g+l+i+n \\ &= h+k+j+m \\ &= h+j+k+m, \end{aligned}$$

as required. □

Let $a = [(k,l)]$ and $b = [(m,n)]$, where k, l, m, n are natural numbers. We define

$$a + b = [(k+m, l+n)].$$

For instance, $0+1 = [(1,1)]+[(2,1)] = [(3,2)] = [(2,1)] = 1$. By Lemma 4.2, if g, h, i, j are natural numbers such that $(g,h) \sim (k,l)$ and $(i,j) \sim (m,n)$ then

$$(g+i, h+j) \sim (k+m, l+n).$$

Thus $(g+i, h+j)$ and $(k+m, l+n)$ determine the same equivalence class $a + b$. Consequently in the definition of $a + b$ it does not matter which representatives (k,l) and (m,n) are selected from the equivalence classes a and b respectively. In other words, $a+b$ denotes a unique equivalence class. We describe this result by asserting that $a+b$ is well defined. We call $a+b$ the **sum** of a and b, and we refer to the binary operation on $\mathbb{Z}$ denoted by $+$ as **addition**.

We move on to properties of addition in $\mathbb{Z}$. In the next two theorems, we obtain the commutativity and associativity of addition of integers easily from the corresponding properties of natural numbers.

Theorem 4.3 *For each $a \in \mathbb{Z}$ and $b \in \mathbb{Z}$ we have $a + b = b + a$.*

Proof. Choose $(k,l) \in a$ and $(m,n) \in b$. Then

$$\begin{aligned} a + b &= [(k+m, l+n)] \\ &= [(m+k, n+l)] \\ &= b + a. \end{aligned}$$
□

Theorem 4.4 *If $(a,b,c) \in \mathbb{Z} \times \mathbb{Z} \times \mathbb{Z}$ then $a + (b+c) = (a+b) + c$.*

Proof. Choose $(i,j) \in a$, $(k,l) \in b$ and $(m,n) \in c$. Then

$$\begin{aligned} a + (b+c) &= [(i,j)] + ([(k,l)] + [(m,n)]) \\ &= [(i,j)] + [(k+m, l+n)] \\ &= [(i+(k+m), j+(l+n))] \\ &= [((i+k)+m, (j+l)+n)] \\ &= [(i+k, j+l)] + [(m,n)] \\ &= ([(i,j)] + [(k,l)]) + [(m,n)] \\ &= (a+b) + c. \end{aligned}$$
□

Theorem 4.5 *We have $a + 0 = a$ for each $a \in \mathbb{Z}$.*

Proof. Choose $(m, n) \in a$. Then
$$\begin{aligned} a + 0 &= [(m, n)] + [(1, 1)] \\ &= [(m + 1, n + 1)] \\ &= [(m, n)] \\ &= a. \end{aligned}$$
□

Theorem 4.6 *For each $a \in \mathbb{Z}$ there exists $b \in \mathbb{Z}$ such that $a + b = 0$.*

Proof. Choose $(m, n) \in a$, and define $b = [(n, m)]$. Then
$$\begin{aligned} a + b &= [(m, n)] + [(n, m)] \\ &= [(m + n, n + m)] \\ &= [(1, 1)] \\ &= 0. \end{aligned}$$
□

4.3 Multiplication

Again we start with a preliminary lemma.

Lemma 4.7 *Let g, h, i, j, k, l, m, n be natural numbers. If $(g, h) \sim (k, l)$ and $(i, j) \sim (m, n)$ then*
$$(gi + hj, hi + gj) \sim (km + ln, lm + kn).$$

Remark: If $k > l$ and $m > n$, then
$$\begin{aligned} km + ln &= (l + (k - l))m + ln \\ &= lm + (k - l)(n + (m - n)) + ln \\ &= lm + ln + (k - l)n + (k - l)(m - n) \\ &= lm + n(l + (k - l)) + (k - l)(m - n) \\ &= lm + kn + (k - l)(m - n). \end{aligned}$$

Hence
$$(k - l)(m - n) = (km + ln) - (lm + kn).$$

This equation explains our interest in the lemma.

Proof. We show first that both $(gi+hj, hi+gj)$ and $(km+ln, lm+kn)$ are equivalent to $(ki+lj, li+kj)$, and then the transitivity of the equivalence relation completes the proof.

We are given that $g + l = h + k$, since $(g, h) \sim (k, l)$. Therefore

$$\begin{aligned} gi + hj + li + kj &= gi + li + hj + kj \\ &= i(g + l) + j(h + k) \\ &= i(h + k) + j(g + l) \\ &= ih + ik + jg + jl \\ &= hi + gj + ki + lj \end{aligned}$$

and so

$$(gi + hj, hi + gj) \sim (ki + lj, li + kj).$$

Since $(m, n) \sim (i, j)$, a symmetrical argument shows that

$$(km + ln, lm + kn) \sim (ki + lj, li + kj),$$

and the lemma follows. □

Let $a = [(k, l)]$ and $b = [(m, n)]$, where k, l, m, n are natural numbers. We define

$$ab = [(km + ln, lm + kn)].$$

Lemma 4.7 guarantees that ab is well defined: it does not matter which ordered pairs $(k, l) \in a$ and $(m, n) \in b$ are chosen in the definition of ab. We call ab the **product** of a and b, and the operation we have just defined is called **multiplication**. We sometimes write $a \cdot b$ instead of ab.

Let us proceed to the elementary properties of multiplication.

Theorem 4.8 *For each $(a, b) \in \mathbb{Z} \times \mathbb{Z}$ we have $ab = ba$.*

Proof. Choose $(k, l) \in a$ and $(m, n) \in b$. Then

$$\begin{aligned} ab &= [(km + ln, lm + kn)] \\ &= [(mk + nl, nk + ml)] \\ &= ba. \end{aligned}$$

□

Theorem 4.9 *If $(a, b, c) \in \mathbb{Z} \times \mathbb{Z} \times \mathbb{Z}$, then $a(bc) = (ab)c$.*

Proof. Choose $(i,j) \in a$, $(k,l) \in b$ and $(m,n) \in c$. Then
$$bc = [(km + ln, lm + kn)],$$
so that
$$\begin{aligned}a(bc) &= [(i(km+ln) + j(lm+kn), j(km+ln) + i(lm+kn))] \\ &= [(ikm + iln + jlm + jkn, jkm + jln + ilm + ikn)].\end{aligned}$$

Similarly
$$\begin{aligned}(ab)c &= c(ab) \\ &= [(mik + mjl + njk + nil, nik + njl + mjk + mil)] \\ &= [(ikm + iln + jlm + jkn, jkm + jln + ilm + ikn)] \\ &= a(bc).\end{aligned}$$
$\square$

Theorem 4.10 *If $(a,b,c) \in \mathbb{Z} \times \mathbb{Z} \times \mathbb{Z}$, then $a(b+c) = ab + ac$.*

Proof. Choose $(i,j) \in a$, $(k,l) \in b$ and $(m,n) \in c$. Then
$$\begin{aligned}a(b+c) &= [(i,j)][(k+m, l+n)] \\ &= [(i(k+m) + j(l+n), j(k+m) + i(l+n))] \\ &= [(ik + im + jl + jn, jk + jm + il + in)]\end{aligned}$$
and
$$\begin{aligned}ab + ac &= [(ik + jl, jk + il)] + [(im + jn, jm + in)] \\ &= [(ik + jl + im + jn, jk + il + jm + in)] \\ &= [(ik + im + jl + jn, jk + jm + il + in)] \\ &= a(b+c).\end{aligned}$$
$\square$

Observe that if $(m,n) \in \mathbb{N} \times \mathbb{N}$ then
$$\begin{aligned}[(m+1,1)] + [(n+1,1)] &= [(m+n+2, 2)] \\ &= [(m+n+1, 1)]\end{aligned}$$
and
$$\begin{aligned}[(m+1,1)][(n+1,1)] &= [((m+1)(n+1) + 1, n+1+m+1)] \\ &= [(m(n+1) + n + 1 + 1, n + m + 2)] \\ &= [(mn + m + n + 2, m + n + 2)] \\ &= [(mn + 1, 1)].\end{aligned}$$

We therefore see that $[(n+1,1)]$ satisfies rules similar to those for n in $\mathbb{N}$. Let us define
$$\Phi(n) = [(n+1,1)]$$
for each $n \in \mathbb{N}$. Then the calculations above show that
$$\Phi(m) + \Phi(n) = \Phi(m+n)$$
and
$$\Phi(m)\Phi(n) = \Phi(mn).$$
We also see that Φ is an injection from $\mathbb{N}$ into $\mathbb{Z}$. Indeed, suppose that $\Phi(m) = \Phi(n)$. Then $[(m+1,1)] = [(n+1,1)]$, so that $(m+1,1) \sim (n+1,1)$. In other words, $m+2 = n+2$, and so $m = n$ by the cancellation law.

The function Φ thus sets up a $1:1$ correspondence between $\mathbb{N}$ and a subset of $\mathbb{Z}$. Moreover, the image of $m+n$ under this correspondence is the sum of the images of m and n. A similar statement holds for products. (Exponentiation will be considered in a more general setting later.) Roughly speaking, what this means is that the integers of the form $[(n+1,1)]$, where $n \in \mathbb{N}$, can be manipulated as if they were natural numbers. For all mathematical purposes, they are indistinguishable from natural numbers, and we shall henceforth identify them with natural numbers. Note also that $\Phi(1) = [(2,1)] = 1$. Thus the two meanings for "1" have been reconciled.

4.4 Rings

At this point it is convenient to generalize. The advantage of doing so is to avoid repeating the same argument in different settings.

A **ring** is an ordered 4-tuple $(R, +, \cdot, 0)$ where R is a set, $0 \in R$ and $+$ and $\cdot$ are binary operations on R satisfying the following axioms:

(a) $a + b = b + a$ for each $(a,b) \in R \times R$;
(b) $a + (b+c) = (a+b) + c$ for each $(a,b,c) \in R \times R \times R$;
(c) $a + 0 = a$ for all $a \in R$;
(d) for each $a \in R$ there exists $b \in R$ such that $a + b = 0$;
(e) $a \cdot (b \cdot c) = (a \cdot b) \cdot c$ for each $(a,b,c) \in R \times R \times R$;
(f) $a \cdot (b+c) = (a \cdot b) + (a \cdot c)$ and $(b+c) \cdot a = (b \cdot a) + (c \cdot a)$ for each $(a,b,c) \in R \times R \times R$.

Clearly we have proved that $(\mathbb{Z}, +, \cdot, 0)$ is a ring.

The operations $+$ and $\cdot$ are called **addition** and **multiplication** respectively. If $a \in R$ and $b \in R$ then we call $a + b$ the **sum** of a and b, and $a \cdot b$ is their **product**. We say that a and b are the **terms** of $a + b$ and the **factors** of $a \cdot b$. We usually write ab instead of $a \cdot b$. Note that $(R, +)$ and $(R, \cdot)$ are semigroups.

Suppose that $a + x = a + y$, where $(a, x, y) \in R \times R \times R$. By (d) there exists $b \in R$ such that $b + a = a + b = 0$. Hence

$$x = 0 + x$$
$$= b + a + x$$
$$= b + a + y$$
$$= 0 + y$$
$$= y.$$

We infer that the cancellation law holds for addition.

The cancellation law implies that 0 is the only element $b \in R$ such that $a + b = a$ for each $a \in R$, for if $a + b = a$ then $a + b = a + 0$, so that $b = 0$ by cancellation. We call 0 the **zero** element, or **additive identity**, of R.

Similarly for each $a \in R$ there is a unique $b \in R$ for which $a + b = 0$. Indeed the existence of b is guaranteed by axiom (d), and if $a + c = 0$ then $a + b = a + c$, so that $c = b$ by cancellation.

We call b the **additive inverse** of a, and denote it by $-a$. Thus

$$a + (-a) = 0.$$

The uniqueness of the additive inverse has important consequences. It implies, for instance, that $-0 = 0$, since $0 + 0 = 0$. Similarly

$$-(-a) = a$$

for each $a \in R$, since $-a + a = 0$. If $(a, b) \in R \times R$, then

$$a + b + (-a) + (-b) = a + (-a) + b + (-b)$$
$$= 0 + 0$$
$$= 0$$

and so

$$-(a + b) = -a + (-b).$$

For each $a \in R$ we also have
$$\begin{aligned} 0 + a \cdot 0 = a \cdot 0 \\ = a(0+0) \\ = a \cdot 0 + a \cdot 0, \end{aligned}$$
so that
$$a \cdot 0 = 0$$
by cancellation. Similarly
$$0 \cdot a = 0.$$
Since
$$\begin{aligned} ab + (-a)b = (a + (-a))b \\ = 0 \cdot b \\ = 0, \end{aligned}$$
the uniqueness of the additive inverse implies that
$$-(ab) = (-a)b.$$
Similarly
$$a(-b) = -(ab).$$
We usually write $-ab$ instead of $-(ab)$, $(-a)b$ or $a(-b)$. We also have
$$\begin{aligned} (-a)(-b) = -(-a)b \\ = -(-ab) \\ = ab. \end{aligned}$$
Suppose $a \in \mathbb{Z}$. Then $a = [(m,n)]$ for some $(m,n) \in \mathbb{N} \times \mathbb{N}$. Since
$$\begin{aligned} [(m,n)] + [(n,m)] = [(m+n, n+m)] \\ = [(1,1)] \\ = 0, \end{aligned}$$
it follows from the uniqueness of the additive inverse that
$$-a = [(n,m)].$$
Recall that any $n \in \mathbb{N}$ can be identified with $[(n+1, 1)]$ by means of the function Φ of the previous section. Thus $-n = [(1, n+1)]$. For example, $-1 = [(1,2)]$.

For any $a \in R$ and $b \in R$ we define
$$a - b = a + (-b).$$
The operation thus denoted by $-$ is called **subtraction**. Note that
$$a - a = 0,$$
and if $a - b = 0$ then the addition of b to both sides of the equation yields $a = b$. Moreover the uniqueness of the additive inverse shows that
$$-(a - b) = b - a,$$
since $a - b + b - a = 0$. Subtraction also satisfies the distributive laws: for each $(a, b, c) \in R \times R \times R$ we have
$$\begin{aligned}a(b - c) &= ab + a(-c) \\ &= ab - ac,\end{aligned}$$
and similarly
$$(a - b)c = ac - bc.$$
Moreover, if $a + b = c$, then the addition of $-a$ to both sides gives
$$b = c - a.$$

A ring is **commutative** if its multiplication is commutative. Thus we see that $(\mathbb{Z}, +, \cdot, 0)$ is a commutative ring.

Let $(R, +, \cdot, 0)$ be a ring. An element $1 \in R - \{0\}$ is called a **unit element** if $a \cdot 1 = 1 \cdot a = a$ for each $a \in R$. If a unit element 1 exists, then $((R, +, \cdot, 0), 1) = (R, +, \cdot, 0, 1)$ is called a **ring with unity**. For example, if $(m, n) \in \mathbb{N} \times \mathbb{N}$ then the definition of multiplication gives
$$\begin{aligned}[(m, n)][(2, 1)] &= [(2m + n, m + 2n)] \\ &= [(m, n)]\end{aligned}$$
since $(m, n) \sim (m+m+n, n+m+n) = (2m+n, m+2n)$. Thus $(\mathbb{Z}, +, \cdot, 0, 1)$ is a ring with unity, $[(2, 1)]$ being a unit element. Moreover $[(2, 1)] = \Phi(1)$ and $[(2, 1)]$ can therefore be identified with 1 in $\mathbb{N}$.

Note that if $(R, +, \cdot, 0, 1)$ is a ring with unity, then 1 is unique. Indeed, if there exists $b \in R$ such that $ba = ab = a$ for each $a \in R$, then $1 = b \cdot 1 = b$.

We also observe that if $a \in R$ then
$$(-1)a = -(1 \cdot a) = -a.$$

Similarly
$$a(-1) = -a.$$

Now suppose that $(R, +, \cdot, 0, 1)$ is a commutative ring with unity, and that $ab \neq 0$ whenever a and b are nonzero elements of R. Then $(R, +, \cdot, 0, 1)$ is called an **integral domain**. We call 0 and 1 the **additive** and **multiplicative identities**, respectively, of the integral domain.

Theorem 4.11 $(\mathbb{Z}, +, \cdot, 0, 1)$ *is an integral domain.*

Proof. Choose nonzero integers a and b. We must show that $ab \neq 0$.

Choose $(k, l) \in a$ and $(m, n) \in b$. Since $a \neq 0$ and $b \neq 0$, it follows that $k \neq l$ and $m \neq n$. We distinguish two cases.

Case I: Suppose $m > n$. Then there exists $j \in \mathbb{N}$ such that $n + j = m$. Since $k \neq l$, we have $kj \neq lj$ by the cancellation law for multiplication in $\mathbb{N}$. Consequently

$$\begin{aligned} km + ln &= k(n + j) + ln \\ &= kn + kj + ln \\ &\neq kn + lj + ln \\ &= kn + l(j + n) \\ &= kn + lm. \end{aligned}$$

Hence

$$\begin{aligned} ab &= [(k, l)][(m, n)] \\ &= [(km + ln, lm + kn)] \\ &\neq 0. \end{aligned}$$

Case II: The argument is similar if $m < n$. $\square$

The next theorem gives a multiplicative cancellation law.

Theorem 4.12 Let $(R, +, \cdot, 0, 1)$ *be an integral domain, and let* $(a, b, c) \in R \times R \times R$. *If* $ab = ac$ *and* $a \neq 0$, *then* $b = c$.

Proof. By adding $-ac$ to (in other words, subtracting ac from) both sides of the equation $ab = ac$, we find that

$$0 = ab - ac = a(b - c).$$

Since $(R, +, \cdot, 0, 1)$ is an integral domain and $a \neq 0$, it follows that $b - c = 0$. Hence $b = c$. $\square$

4.5 Exponentiation

Let $(R, +, \cdot, 0, 1)$ be an integral domain. For each $a \in R$ and $n \in \mathbb{N}$ we define

$$a^n = \prod_{k=1}^{n} a.$$

We also define $a^0 = 1$ if $a \neq 0$.

If $(a, n) \in \mathbb{N} \times \mathbb{N}$, then the notation above needs to be reconciled with that used earlier. This reconciliation is accomplished by induction on n. First, $a^1 = \prod_{k=1}^{1} a = a$. Next, assume as an inductive hypothesis that $\prod_{k=1}^{n} a = a^n$ for some $n \in \mathbb{N}$ (where we are using our previous definition of a^n). Then

$$\prod_{k=1}^{n+1} a = a \prod_{k=1}^{n} a = a \cdot a^n = a^{n+1},$$

as required.

Note that, in general, if $(a, n) \in R \times \mathbb{N}$ then

$$a^{n+1} = \prod_{k=1}^{n+1} a = a \prod_{k=1}^{n} a = a \cdot a^n.$$

We refer to a^n as the nth **power** of a. We call a^2 and a^3 the **square** and **cube**, respectively, of a. If $(a, b) \in R \times R$, note that

$$\begin{aligned}(a+b)^2 &= (a+b)(a+b) \\ &= a(a+b) + b(a+b) \\ &= a^2 + ab + ba + b^2 \\ &= a^2 + 2ab + b^2.\end{aligned}$$

Similarly

$$\begin{aligned}(a-b)(a+b) &= a(a+b) + (-b)(a+b) \\ &= a^2 + ab - ba - b^2 \\ &= a^2 - b^2.\end{aligned}$$

Theorem 4.13 *Let $(R, +, \cdot, 0, 1)$ be an integral domain. Let $(a, b) \in R \times R$ and let $n \in \mathbb{N}$. Then*

$$(ab)^n = a^n b^n.$$

Proof. Both sides are equal to ab if $n = 1$. Suppose as an inductive hypothesis that $(ab)^n = a^n b^n$ for some $n \in \mathbb{N}$. Then

$$\begin{aligned}(ab)^{n+1} &= ab(ab)^n \\ &= aba^n b^n \\ &= (a \cdot a^n)(b \cdot b^n) \\ &= a^{n+1} b^{n+1},\end{aligned}$$

as required. □

The next theorem is a generalization of Theorems 3.22 and 3.23. Its proof is identical to the proofs of those theorems, and is therefore omitted.

Theorem 4.14 *Let $(R, +, \cdot, 0, 1)$ be an integral domain, and let $(a, m, n) \in R \times \mathbb{N} \times \mathbb{N}$. Then:*

(a) $a^m a^n = a^{m+n}$;
(b) $(a^m)^n = a^{mn}$.

4.6 Order

We proceed to extend to integers the idea of order that was previously introduced for natural numbers. We begin with the following lemma.

Lemma 4.15 *Let $a \in \mathbb{Z}$, $(k, l) \in a$ and $(m, n) \in a$. Then $k < l$ if and only if $m < n$.*

Proof. Since $(k, l) \in a$ and $(m, n) \in a$, we have $(k, l) \sim (m, n)$, and so $k + n = l + m$. Suppose $k < l$. If $n \leq m$, then we should obtain the contradiction that $k + n < l + m$. Hence $m < n$. Similarly if $m < n$ then $k < l$. □

Of course it is also true that $k > l$ if and only if $m > n$.

Let $a \in \mathbb{Z}$ and $(m, n) \in a$. We say that a is **positive** if $m > n$ and **negative** if $m < n$. Lemma 4.15 assures us that these concepts are well defined: it does not matter which representative of the equivalence class a is selected. Note that 0 is neither positive nor negative, since $(1, 1) \in 0$ and $1 \not< 1$. On the other hand, 1 is positive since $(2, 1) \in 1$ and $2 > 1$. Any nonzero integer is either positive or negative, but not both.

Let P be the set of positive integers. Choose $a \in P$ and $(m, n) \in a$. Then $m > n$, and so by definition we may write $m = n + k$ for some $k \in \mathbb{N}$. Thus $(n + k, n) \in a$, so that $(k + 1, 1) \in a$ since $(n + k, n) \sim$

$(k+1, 1)$ by definition. Therefore $a \in \mathbb{N}$ by our earlier identification of the natural numbers with the integers of the form $[(n+1,1)]$ where $n \in \mathbb{N}$. Conversely each natural number is positive, since $n+1 > 1$ for each $n \in \mathbb{N}$. Thus we can now identify the natural numbers with the positive integers. One consequence is that sums and products of positive integers are always positive. We also have the following lemma.

Lemma 4.16 *Let $a \in \mathbb{Z} - \{0\}$. Then either a or $-a$ is positive, but not both.*

Proof. Choose $(m, n) \in a$. Since $a \neq 0$, we have $m \neq n$. If $m > n$ then a is positive. Otherwise $m < n$. In this case $-a$ is positive, since $-a = [(n, m)]$. However a and $-a$ are not both positive, since $a + (-a) = 0$, which is not positive. □

One consequence of Lemma 4.16 is that $-a$ is positive if and only if a is negative, for if $a = 0$ then $-a = 0$.

Let $(a, b) \in \mathbb{Z} \times \mathbb{Z}$. We write $a < b$, or $b > a$, if $b - a$ is positive. In this case we say that a is **less** than b, or that b is **greater** than a. Since $b - a$ is positive, there exists $n \in \mathbb{N}$ such that $b - a = n$. Therefore $a + n = b$ (by the addition of a to both sides of the equation). Thus if a and b are positive then the current notation and terminology are reconciled with those used previously for natural numbers.

It follows immediately from the above definitions that an integer $a = a - 0$ is positive if and only if $a > 0$. Similarly a is negative if and only if $a < 0$. If $a \neq 0$ then we say that the **sign** of a is positive or negative according to whether $a > 0$ or $a < 0$.

We move on to the elementary properties of the notion of order we have just defined. The symbol "$<$" defines a relation on $\mathbb{Z}$. If $a \in \mathbb{Z}$ then $a \not< a$, for $a - a = 0$, which is not positive. This relation is therefore not reflexive. It is, however, transitive. Indeed, suppose $(a, b, c) \in \mathbb{Z} \times \mathbb{Z} \times \mathbb{Z}$, and that $a < b$ and $b < c$. Then $b - a > 0$ and $c - b > 0$. Hence $c - a = c - b + b - a > 0$, so that $a < c$.

Recall that if $(a, b) \in \mathbb{Z} \times \mathbb{Z}$ and $b - a \neq 0$ then either $b - a$ or $-(b - a)$ is positive. If $b - a > 0$ then $a < b$ by definition. On the other hand, if $-(b - a) = a - b > 0$, then $a > b$. Of course if $b - a = 0$ then $a = b$. Thus one of the possibilities $a = b$, $a < b$ and $a > b$ must hold.

Theorem 4.17 *Let $(a, b, c) \in \mathbb{Z} \times \mathbb{Z} \times \mathbb{Z}$. If $a < b$ then $a + c < b + c$.*

Proof. Since $a < b$, $b - a$ is positive. Hence
$$b + c - (a + c) = b + c - a - c$$
$$= b - a$$
$$> 0,$$
and so $a + c < b + c$. □

Corollary 4.18 *Let $(a, b) \in \mathbb{Z} \times \mathbb{Z}$. If $a < b$ then $a + 1 \leq b$.*

Proof. Otherwise $a < b < a + 1$. Subtracting a, we obtain $0 < b - a < 1$. Since $b - a > 0$, we have $b - a \in \mathbb{N}$. But no natural number is less than 1. This contradiction implies that $a + 1 \leq b$. □

Theorem 4.19 *Let $(a, b, c) \in \mathbb{Z} \times \mathbb{Z} \times \mathbb{Z}$. If $a < b$ and $c > 0$ then $ac < bc$.*

Remark: What this lemma says is that both sides of an inequality may be multiplied by a *positive* integer.

Proof. Since $a < b$, we have $b - a > 0$. Hence
$$bc - ac = c(b - a) > 0$$
since $c > 0$, and so $bc > ac$. □

Once again we proceed to generalize. Let $(R, +, \cdot, 0, 1)$ be an integral domain, and suppose that $<$ is a relation on R satisfying the following axioms for each $(a, b, c) \in R \times R \times R$:

(a) $a \not< a$;
(b) if $a < b$ and $b < c$, then $a < c$;
(c) $a = b$, $a < b$ or $a > b$;
(d) if $a < b$ then $a + c < b + c$;
(e) if $a < b$ and $c > 0$, then $ac < bc$.

We call $((R, +, \cdot, 0, 1), <) = (R, +, \cdot, 0, 1, <)$ an **ordered integral domain**. Thus $(\mathbb{Z}, +, \cdot, 0, 1, <)$ is an ordered integral domain. If $(a, b) \in R \times R$ and $a < b$, then we say that a is **less** than b, and that b is **greater** than a. Elements of R that are greater than 0 are said to be **positive**; those less than 0 are **negative**. We proceed to investigate some elementary properties of ordered integral domains.

Note first that, as in the case of natural numbers, axioms (a) and (b) imply that each of the possibilities listed in (c) rules out the others. In other words, exactly one of the possibilities in (c) must hold.

For each $(a,b) \in R \times R$ we write $a \leq b$ or $b \geq a$ if either $a < b$ or $a = b$. Thus $a \leq a$. Moreover, if $a < b$ and $b \leq c$, or $a \leq b$ and $b < c$, then $a < c$, and the transitivity of the relation denoted by $\leq$ follows as in the case of natural numbers. Furthermore, if $a \leq b \leq a$, then $a = b$, for otherwise $a < a$ by transitivity. Again as with natural numbers we have $a + c < b + d$ if a, b, c, d are elements of R such that $a < b$ and $c < d$. For example, if $b > 0$ and $d > 0$ then $b + d > 0$, and if $a < 0$ and $c < 0$ then $a + c < 0$. The inequality $a + c < b + d$ also holds if $a \leq b$ and $c < d$, or if $a < b$ and $c \leq d$, whereas $a + c \leq b + d$ if $a \leq b$ and $c \leq d$. Similarly if $0 \leq a < b$ and $0 \leq c < d$, or if $0 < a \leq b$ and $0 < c < d$, or if $0 < a < b$ and $0 < c \leq d$, then $ac < bd$, whereas if $0 \leq a \leq b$ and $0 \leq c \leq d$ then $ac \leq bd$. If $ab < ac$ and $a > 0$, then $b < c$, as the other possibilities lead to immediate contradictions.

Suppose $a < b$. Adding $-a$ to both sides, we obtain $b - a > 0$. Conversely if $b - a > 0$, then we find that $a < b$ by adding a to both sides. Similarly $a < b$ if and only if $a - b < 0$. In particular, $a > 0$ if and only if $-a < 0$, and $a < 0$ if and only if $-a > 0$.

The next theorem deals with the multiplication of both sides of an inequality by a negative element.

Theorem 4.20 Let $(R, +, \cdot, 0, 1, <)$ be an ordered integral domain and let $(a, b, c) \in R \times R \times R$. If $a < b$ and $c < 0$, then $ac > bc$.

Remark: Multiplication of both sides of an inequality by a negative element of R therefore causes a change in the direction of the inequality. For example, $1 < 2$ but $-1 > -2$.

Proof. Since $c < 0$, we have $-c > 0$. Therefore $-ac < -bc$ since $a < b$. Thus
$$-(bc - ac) = -bc - (-ac) > 0,$$
so that $bc - ac < 0$. Hence $bc < ac$, as required. $\square$

Thus if $c \neq 0$ and $ac = bc$ then $a = b$, as the other possibilities lead to immediate contradictions by the theorem above or axiom (e). This result can be regarded as a cancellation law for multiplication.

We next investigate the signs of products of elements of an integral domain. If $a > 0$ and $b > 0$, then $ab > 0$ by axiom (e). However $a(-b) = -ab < 0$, and $(-a)(-b) = ab > 0$. Thus the product of two nonzero elements is negative if and only if exactly one of the factors is negative. It follows that $a^2 > 0$ if $a \neq 0$, and therefore that $a^2 \geq 0$ in any case.

Finally, let X be a non-empty finite subset of R. We write $\max X$ and $\min X$ respectively for the greatest and least elements of X. Thus if $X = \{x_1, x_2, \ldots, x_n\}$, where $n > 1$, then

$$\max X = \max\{\max(X - \{x_n\}), x_n\},$$

and a corresponding result holds for $\min X$.

4.7 Square Roots and Absolute Values

It is convenient here to introduce the notion of the absolute value of a member of an ordered integral domain. This concept will be used extensively in Chapter 4.9.

We begin with the idea of a square root of a member of an ordered integral domain $(R, +, \cdot, 0, 1, <)$. If $(a, b) \in R \times R$ and $b^2 = a$, then we call b a **square root** of a. For example, 0 is a square root of 0, since $0^2 = 0$. Since $(-b)^2 = b^2 = a$, we see that $-b$ is also a square root of a if b is. On the other hand, suppose b and c are square roots of a. Then $b^2 = c^2 = a$. Hence

$$0 = b^2 - c^2 = (b+c)(b-c).$$

Thus either $b + c = 0$ or $b - c = 0$, and so $b = c$ or $b = -c$. We deduce that 0 is the only square root of 0. If $a \neq 0$ and a has a square root, then a has exactly two square roots, one positive and the other negative. In summary, if a has a square root, then $a \geq 0$ (since $b^2 \geq 0$ for any $b \in R$), and a has a unique non-negative square root, which we denote by $\sqrt{a}$. Another way of stating the latter result is that if $b \geq 0$, $c \geq 0$ and $b^2 = c^2$, then $b = c$. However if $b^2 < c^2$ for non-negative integers b and c, then $b < c$ as the other possibilities lead to contradictions.

Note that if b and c are elements of R such that $\sqrt{b}$ and $\sqrt{c}$ exist, then

$$\left(\sqrt{b} \cdot \sqrt{c}\right)^2 = \left(\sqrt{b}\right)^2 \cdot \left(\sqrt{c}\right)^2 = bc$$

and

$$\sqrt{b} \cdot \sqrt{c} \geq 0.$$

Hence

$$\sqrt{b} \cdot \sqrt{c} = \sqrt{bc}$$

by the definition of $\sqrt{bc}$.

The **absolute value**, $|a|$, of an element a of R is defined as $\sqrt{a^2}$. Note that the only square roots of a^2 are a and $-a$. Since $\sqrt{a^2} \geq 0$, we have $|a| = a$ if $a \geq 0$ and $|a| = -a$ otherwise. Thus, for example, $|5| = 5$, $|-3| = 3$ and $|0| = 0$. Note that $|a| \geq 0$,

$$|-a| = \sqrt{(-a)^2} = \sqrt{a^2} = |a|,$$

$a \leq |a|$, and $-a \leq |a|$. Thus $-|a| \leq a \leq |a|$. Moreover, if $|a| = |b|$ where $(a,b) \in R \times R$, then $\sqrt{a^2} = \sqrt{b^2}$. Hence $a^2 = b^2$, so that $a = \pm b$. (By this notation we mean that $a \in \{b, -b\}$.) This argument is reversible: if $a = \pm b$ then $|a| = |b|$. Thus if $b \geq 0$ then $|a| = b$ if and only if $a = \pm b$. For example, $|a| = 0$ if and only if $a = 0$.

The following theorem is particularly important.

Theorem 4.21 *Let $(R, +, \cdot, 0, 1, <)$ be an ordered integral domain, and let $(a,b) \in R \times R$ where $b > 0$. Then $|a| < b$ if and only if $-b < a < b$.*

Proof. Suppose $|a| < b$. Then by definition $\sqrt{a^2} < b$. Squaring, we obtain $a^2 < b^2$, and so

$$0 > a^2 - b^2 = (a-b)(a+b).$$

Therefore one of $a - b$ and $a + b$ is negative and the other positive. Since $b > 0 > -b$, we have $a - b < a + b$, and so $a - b < 0$ and $a + b > 0$. Therefore $-b < a < b$.

The argument above is reversible. Thus if $-b < a < b$, then $|a| < b$, and the proof is complete. □

Similarly $|a| \leq b$ if and only if $-b \leq a \leq b$.

Corollary 4.22 *If $(a,b,c) \in R \times R \times R$ and $b > 0$, then $|a - c| < b$ if and only if $c - b < a < c + b$.*

Proof. By Theorem 4.21 we have $|a-c| < b$ if and only if $-b < a - c < b$. The result follows by adding c throughout the latter inequalities. □

Similarly we have $|a - c| \leq b$ if and only if $c - b \leq a \leq c + b$. It also follows from Theorem 4.21 that $|a| \geq b$ if and only if $a \leq -b$ or $a \geq b$. Similarly $|a| > b$ if and only if $a < -b$ or $a > b$.

Next, let $(a,b) \in R \times R$. Then

$$|ab| = \sqrt{(ab)^2} = \sqrt{a^2 b^2} = \sqrt{a^2} \cdot \sqrt{b^2} = |a||b|.$$

Although, as we have just seen, the absolute value of a product is equal to the product of the absolute values of the factors, the corresponding assertion for sums is not true. Instead we have the following theorem, which is known as the triangle inequality because of its analogy to the geometric fact that the sum of the lengths of any two sides of a triangle is greater than the length of the third side.

Theorem 4.23 (**Triangle Inequality**) *If $(R, +, \cdot, 0, 1, <)$ is an ordered integral domain and $(a, b) \in R \times R$ then $|a + b| \leq |a| + |b|$.*

Proof. We have $-|a| \leq a \leq |a|$ and $-|b| \leq b \leq |b|$. Adding, we obtain

$$-(|a| + |b|) \leq a + b \leq |a| + |b|,$$

and so $|a + b| \leq |a| + |b|$ by Theorem 4.21. $\square$

Note also that

$$\begin{aligned}|a - b| &= |a + (-b)| \\ &\leq |a| + |-b| \\ &= |a| + |b|.\end{aligned}$$

In addition, $|a| = |a + b - b| \leq |a + b| + |b|$, so that $|a| - |b| \leq |a + b|$. Similarly, $|b| - |a| \leq |a + b|$, and so

$$||a| - |b|| \leq |a + b|.$$

Since $|-b| = |b|$ it also follows that $||a| - |b|| \leq |a - b|$.

4.8 An Extension of Induction

We have seen that induction provides a powerful tool for proving theorems about the natural numbers. The natural numbers can be viewed as those integers that are no less than 1. For any integer a we shall now extend the notion of induction so as to supply a method for proving theorems about integers no less than a. We will then be able to start the inductive process at any integer, whereas we previously started it at 1. The proof of the following theorem is essentially just an application of Theorem 3.31.

Theorem 4.24 *Let $Y \subseteq \mathbb{Z}$ and $a \in \mathbb{Z}$. Suppose that Y contains every integer b such that $c \in Y$ for each integer c satisfying $a \leq c < b$. Then Y contains every integer greater than a.*

Remark: Note that the hypothesis concerning Y implies vacuously that $a \in Y$.

Proof. Choose an integer $d \geq a$. Let Y be the set of all integers b such that $c \in Y$ for each integer c satisfying $a \leq c < b$. We must show that $d \in Y$.

Since $d \geq a$, we have $d - a + 1 \in \mathbb{N}$. Let M be the set of all $n \in \mathbb{N}$ such that $n + a - 1 \in Y$. We shall show that $M = \mathbb{N}$. Choose $n \in \mathbb{N}$ and suppose that $m \in M$ for each natural number $m < n$. We must prove that $n \in M$, that is, that $n + a - 1 \in Y$. Choose an integer c such that $a \leq c < n + a - 1$. It suffices to show that $c \in Y$. But $c = m + a - 1$ for some integer m such that $1 \leq m < n$. Thus $m \in M$ by hypothesis, so that $m + a - 1 \in Y$ by the definition of M, as required. Hence $n \in M$, and so $M = \mathbb{N}$ by Theorem 3.31. Thus $n + a - 1 \in Y$ for each $n \in \mathbb{N}$. Since $d - a + 1 = n$ for some $n \in \mathbb{N}$, we have $d = n + a - 1 \in Y$, as desired. □

Corollary 4.25 *Let $Y \subseteq \mathbb{Z}$ and $a \in Y$. Suppose also that $d + 1 \in Y$ whenever $d \in Y$. Then Y contains every integer greater than a.*

Proof. Choose an integer $b > a$. We must show that $b \in Y$. By Theorem 4.24 we may suppose that $c \in Y$ for each integer c satisfying $a \leq c < b$. By the hypothesis of the corollary, it suffices to show that $b - 1 \in Y$. But $a \leq b - 1$ since $a < b$, and clearly $b - 1 < b$. By the hypothesis on c it follows that $b - 1 \in Y$, as required. □

This corollary is the extension to integers of the version of induction given in Theorem 3.2. Once again it enables us to start the induction at an arbitrary integer, rather than the natural number 1.

4.9 Sums and Products

We proceed to extend the ideas of Section 3.6. The results we establish are valid for any semigroup. Recall that addition in a semigroup is associative. The associative law can be generalized, as in the following theorem. The idea is that before an addition is performed, the summands may be partitioned into sets in any way we please without affecting the result of the calculation, provided only that the summands in each of the resulting sets are consecutive. The number of these sets is at most the number of summands. The summands in each set are then added separately, and the resulting sums are added together to yield the desired total. In the statement of the theorem below, F is a function which determines how the summands are apportioned into h sets: it specifies the last summand in each set.

Recall that if m and n are natural numbers such that $m \leq n$, then $[m, n]$ has been defined as the set of all natural numbers k such that $m \leq k \leq n$. More generally, if $(a, b) \in \mathbb{Z} \times \mathbb{Z}$ and $a \leq b$ then we define $[a, b]$ to be the set of all integers c such that $a \leq c \leq b$.

Theorem 4.26 Let $(X, +)$ be a semigroup, and G a function from $\mathbb{N}$ into X. Let $(m, n) \in \mathbb{N} \times \mathbb{N}$, where $m < n$, and let $h \in [1, n - m + 1]$. Let F be a function from $[1, h]$ into $[m, n]$ such that $F(h) = n$ and $F(i) < F(j)$ whenever $1 \leq i < j \leq h$. Define

$$Q(1) = \sum_{k=m}^{F(1)} G(k)$$

and

$$Q(j) = \sum_{k=F(j-1)+1}^{F(j)} G(k)$$

for each $j \in \mathbb{N}$ such that $1 < j \leq h$. Then

$$\sum_{k=m}^{n} G(k) = \sum_{j=1}^{h} Q(j).$$

Remark: It can be shown that the function F always exists. For example, define $F(j) = m + j - 1$ for each $j \in \mathbb{N}$ such that $j < h$. Then $F(j) \geq m$, since $j \geq 1$, and

$$F(j) < m + h - 1 \leq m + n - m + 1 - 1 = n = F(h).$$

Moreover $F(i) < F(j)$ if $i < j < h$, where $i \in \mathbb{N}$. The theorem enables us to group together the terms in a sum in any way we please.

Proof. By Theorem 3.31 we may assume as an inductive hypothesis that the theorem holds for each natural number less than n.

If $h = 1$, then $F(1) = F(h) = n$, and so

$$\sum_{k=m}^{n} G(k) = \sum_{k=m}^{F(1)} G(k) = Q(1) = \sum_{j=1}^{h} Q(j).$$

Suppose therefore that $h > 1$. Then

$$\sum_{j=1}^{h} Q(j) = \sum_{j=1}^{h-1} Q(j) + Q(h)$$

$$= \sum_{k=m}^{F(h-1)} G(k) + \sum_{k=F(h-1)+1}^{n} G(k), \quad (4.1)$$

since the theorem is assumed to hold with n replaced by $F(h-1) < n$. If $F(h-1) = n-1$, then it follows that

$$\sum_{j=1}^{h} Q(j) = \sum_{k=m}^{n-1} G(k) + \sum_{k=n}^{n} G(k)$$

$$= \sum_{k=m}^{n-1} G(k) + G(n)$$

$$= \sum_{k=m}^{n} G(k),$$

as required.

Suppose therefore that $F(h-1) < n-1$. Then

$$F(h-1) + 1 \leq n-1,$$

and so from 4.1 we obtain

$$\sum_{j=1}^{h} Q(j) = \sum_{k=m}^{F(h-1)} G(k) + \left(\sum_{k=F(h-1)+1}^{n-1} G(k) + G(n) \right)$$

$$= \left(\sum_{k=m}^{F(h-1)} G(k) + \sum_{k=F(h-1)+1}^{n-1} G(k) \right) + G(n).$$

By the inductive hypothesis we may apply the theorem with n replaced by $n-1$, h by 2, $F(1)$ by $F(h-1)$ and $F(2)$ by $n-1$. We therefore deduce that

$$\sum_{j=1}^{h} Q(j) = \sum_{k=m}^{n-1} G(k) + G(n) = \sum_{k=m}^{n} G(k),$$

and the proof is complete by induction. □

For example, let r be a natural number such that $m \leq r < n$. Taking $h = 2$, $F(1) = r$ and $F(2) = n$ gives

$$\sum_{k=m}^{n} G(k) = \sum_{k=m}^{r} G(k) + \sum_{k=r+1}^{n} G(k).$$

Theorem 4.26 justifies the omission of parentheses when a sum is written in full. Thus we may write

$$\sum_{k=m}^{n} G(k) = G(m) + G(m+1) + \cdots + G(n).$$

For each natural number $j \leq n - m + 1$ we may speak of $G(m + j - 1)$ as the jth **term** of the sum.

We shall now extend the idea of a summation so as to permit m and n to be any integers such that $m \leq n$. First, let c be any fixed integer greater than $-m$. Thus $m + c > 0$. It then follows that

$$\sum_{k=m+c}^{n+c} G(k - c) = \sum_{k=m}^{n} G(k), \tag{4.2}$$

both these sums being equal to

$$G(m) + G(m+1) + \cdots + G(n).$$

(More formally, we observe that the jth term of each sum is $G(m + j - 1)$ for every natural number $j \leq n - m + 1$.) Note that the sum on the left of 4.2 is therefore independent of c.

This result suggests the following definition. Let a and b be integers such that $a \leq 0$ and $a \leq b$. Then $-a \geq 0$, so that $1 - a > 0$. Writing $c = 1 - 2a$, we conclude that $a + c = 1 - a > 0$. Moreover $a + c \leq b + c$. Thus if G is a function from $\mathbb{Z}$ into X then we can evaluate

$$\sum_{k=a+c}^{b+c} G(k - c).$$

Motivated by 4.2, we therefore define

$$\sum_{k=a}^{b} G(k) = \sum_{k=1-a}^{b-2a+1} G(k + 2a - 1),$$

where a and b are integers such that $a \leq 0$ and $a \leq b$. Thus we now have $\sum_{k=a}^{b} G(k)$ defined whenever a and b are integers such that $a \leq b$, and we write

$$\sum_{k=a}^{b} G(k) = G(a) + G(a+1) + \cdots + G(b).$$

Let $(X, +)$ be a semigroup. Suppose there exists an element $0 \in X$ such that $x + 0 = 0 + x = x$ for each $x \in X$. Then the ordered pair $((X, +), 0) = (X, +, 0)$ is called a **monoid**. We call 0 the **identity** element of the monoid. For example, $(\mathbb{Z}, +, 0)$ and $(\mathbb{Z}, \cdot, 1)$ are monoids with identity elements 0 and 1 respectively.

Let $(X, +, 0)$ be a monoid and G a function from $\mathbb{Z}$ into X. If a is an integer, then we define

$$\sum_{k=a}^{a-1} G(k) = 0,$$

where 0 is of course the identity element of the monoid. Thus, for example, if G is a function from $\mathbb{Z}$ into $\mathbb{Z}$ then

$$\prod_{k=a}^{a-1} G(k) = 1.$$

As a particular case we have $\prod_{k=1}^{0} k = 1$. This product is denoted by $0!$.

We can also prove a general commutative law, which asserts that terms in a sum may be added in any order. In the following theorem, the function F governs the order in which the terms are to be added.

Theorem 4.27 *Let $(X, +)$ be a commutative semigroup, and let G be a function from $\mathbb{Z}$ into X. Let $(a, b) \in \mathbb{Z} \times \mathbb{Z}$, where $a \leq b$, and let F be a bijection from $[a, b]$ onto itself. Then*

$$\sum_{k=a}^{b} G(k) = \sum_{k=a}^{b} G(F(k)).$$

Proof. We use induction on b. If $b = a$, then we must have $F(a) = a$ since $[a, b] = \{a\}$, and the theorem holds, because

$$\sum_{k=a}^{a} G(k) = G(a) = G(F(a)) = \sum_{k=a}^{a} G(F(k)).$$

We therefore suppose as an inductive hypothesis that the theorem holds for b, and we prove that it holds with b replaced by $b+1$. Assume therefore that F is a bijection from $[a, b+1]$ onto itself. If $F(b+1) = b+1$, then

$$\sum_{k=a}^{b+1} G(k) = \sum_{k=a}^{b} G(k) + G(b+1)$$
$$= \sum_{k=a}^{b} G(F(k)) + G(F(b+1))$$
$$= \sum_{k=a}^{b+1} G(F(k)),$$

as required.

Suppose therefore that $F(c) = b+1$, where $a \le c \le b$. By Theorem 4.26 we obtain

$$\sum_{k=a}^{b+1} G(F(k)) = \sum_{k=a}^{c-1} G(F(k)) + G(F(c)) + \sum_{k=c+1}^{b+1} G(F(k))$$
$$= \sum_{k=a}^{c-1} G(F(k)) + \sum_{k=c}^{b} G(F(k+1)) + G(b+1)$$
$$= \sum_{k=a}^{b} G(H(k)) + G(b+1)$$

where $H(k) = F(k)$ if $k < c$ and $H(k) = F(k+1)$ otherwise. Thus H is an injection, since F is. Since F is a surjection from $[a, b+1]$ onto itself and $F(c) = b+1$, it follows that H is a bijection from $[a, b]$ onto itself. Thus we can apply the inductive hypothesis to conclude that

$$\sum_{k=a}^{b} G(H(k)) = \sum_{k=a}^{b} G(k).$$

Hence

$$\sum_{k=a}^{b+1} G(F(k)) = \sum_{k=a}^{b} G(k) + G(b+1) = \sum_{k=a}^{b+1} G(k),$$

as required. □

In particular, by taking $F(k) = b - k + a$ for all $k \in [a, b]$ we obtain

$$\sum_{k=a}^{b} G(k) = \sum_{k=a}^{b} G(b - k + a).$$

We continue with further elementary properties of the summation notation.

Theorem 4.28 *Let $(R, +, \cdot, 0)$ be a ring, and let G and H be functions from $\mathbb{Z}$ into R. Let $(a, b) \in \mathbb{Z} \times \mathbb{Z}$, where $a \leq b$, and let $(c, d) \in R \times R$. Then*

$$\sum_{k=a}^{b}(cG(k) + dH(k)) = c\sum_{k=a}^{b} G(k) + d\sum_{k=a}^{b} H(k).$$

Proof. The proof is by induction on b. For $b = a$ we have

$$\sum_{k=a}^{a}(cG(k) + dH(k)) = cG(a) + dH(a)$$
$$= c\sum_{k=a}^{a} G(k) + d\sum_{k=a}^{a} H(k).$$

Now suppose that

$$\sum_{k=a}^{b}(cG(k) + dH(k)) = c\sum_{k=a}^{b} G(k) + d\sum_{k=a}^{b} H(k)$$

where a and b are integers such that $b \geq a$. Then

$$\sum_{k=a}^{b+1}(cG(k) + dH(k)) = \sum_{k=a}^{b}(cG(k) + dH(k)) + cG(b+1) + dH(b+1)$$
$$= c\sum_{k=a}^{b} G(k) + d\sum_{k=a}^{b} H(k) + cG(b+1) + dH(b+1)$$
$$= c\left(\sum_{k=a}^{b} G(k) + G(b+1)\right) + d\left(\sum_{k=a}^{b} H(k) + H(b+1)\right)$$
$$= c\sum_{k=a}^{b+1} G(k) + d\sum_{k=a}^{b+1} H(k),$$

and the proof by induction is complete. □

For example, by taking $d = 0$ we obtain

$$\sum_{k=a}^{b} cG(k) = c\sum_{k=a}^{b} G(k).$$

This result serves as a generalized distributive law. Similarly if $(R, +, \cdot, 0, 1)$ is a ring with unity then by putting $c = d = 1$ we find that

$$\sum_{k=a}^{b}(G(k) + H(k)) = \sum_{k=a}^{b} G(k) + \sum_{k=a}^{b} H(k),$$

and by setting $c = 1$ and $d = -1$ we have

$$\sum_{k=a}^{b}(G(k) - H(k)) = \sum_{k=a}^{b} G(k) - \sum_{k=a}^{b} H(k).$$

Thus

$$\sum_{k=a}^{b}(G(k+1) - G(k)) = \sum_{k=a}^{b} G(k+1) - \sum_{k=a}^{b} G(k)$$

$$= \sum_{k=a+1}^{b+1} G(k) - \sum_{k=a}^{b} G(k)$$

$$= \sum_{k=a+1}^{b} G(k) + G(b+1) - \left(\sum_{k=a+1}^{b} G(k) + G(a)\right)$$

$$= G(b+1) - G(a),$$

a result known as the **telescoping property**. For example, if $G(k) = k$ for each k we obtain

$$\sum_{k=a}^{b} 1 = \sum_{k=a}^{b}(k+1-k) = b+1-a$$

by the telescoping property. In particular, if $a = 1$ and b is a natural number n, then

$$\sum_{k=1}^{n} 1 = n + 1 - 1 = n,$$

in agreement with a result obtained earlier.

Example 4.1 For a less trivial application of the telescoping property, let us evaluate $\sum_{k=1}^{n}(2k+1)$, where $n \in \mathbb{N}$. Note first that

$$(k+1)^2 - k^2 = k^2 + 2k + 1 - k^2$$
$$= 2k + 1.$$

Evidently in the telescoping property we must take $G(k) = k^2$ for all k. The telescoping property gives

$$\sum_{k=1}^{n}(2k+1) = \sum_{k=1}^{n}((k+1)^2 - k^2)$$
$$= (n+1)^2 - 1$$
$$= n^2 + 2n + 1 - 1$$
$$= n^2 + 2n.$$

But we also have

$$\sum_{k=1}^{n}(2k+1) = 2\sum_{k=1}^{n}k + \sum_{k=1}^{n}1$$
$$= 2\sum_{k=1}^{n}k + n.$$

Therefore

$$2\sum_{k=1}^{n}k = \sum_{k=1}^{n}(2k+1) - n$$
$$= n^2 + 2n - n$$
$$= n^2 + n$$
$$= n(n+1).$$

△

We also obtain the following theorem.

Theorem 4.29 *If $(a, b, n) \in R \times R \times \mathbb{N}$ for some ring R, then*

$$(a-b)\sum_{k=0}^{n-1} a^k b^{n-k-1} = a^n - b^n,$$

where the convention that $0^0 = 1$ is used.

Proof. Using distributivity we obtain

$$(a-b)\sum_{k=0}^{n-1} a^k b^{n-k-1} = \sum_{k=0}^{n-1}(a^{k+1}b^{n-k-1} - a^k b^{n-k})$$
$$= a^n - b^n$$

by the telescoping property. □

Corollary 4.30 *If R is a ring with unity and $(a, n) \in R \times \mathbb{N}$ then*

$$(a - 1) \sum_{k=0}^{n-1} a^k = a^n - 1,$$

where $0^0 = 1$.

We turn now to a convenient representation of positive integers. We begin with a result called the division theorem.

Theorem 4.31 (**Division Theorem**) *Let $(a, b) \in \mathbb{Z} \times \mathbb{N}$. Then there exist unique integers q and r such that $a = bq + r$ and $0 \leq r < b$. Moreover if $a > 0$ then $q \geq 0$, if $a = 0$ then $q = 0$ and if $a < 0$ then $q \leq 0$.*

Proof. If $b = 1$ then $q = a$ and $r = 0$. We therefore assume that $b > 1$. We prove the existence of q and r first, beginning with the case where $a \geq 0$.

Let X be the set of natural numbers n such that $a < bn$. For example, if $a = 0$ then $1 \in X$. If $a > 0$ we have $ba > a$ since $b > 1$, and so $a \in X$. Hence $X \neq \emptyset$ in both cases. Consequently X has a least element $m \geq 1$. Let $q = m - 1 \geq 0$. If $q = 0$ then $bq = 0 \leq a$. Otherwise $q \in \mathbb{N}$. In this case, since m is the least element of X we have $q \notin X$, and so in both cases $bq \leq a < bm$ since $m \in X$. The case where $a = bq$ satisfies the theorem with $r = 0$ (if $a = 0$ then $q = 0$ since $b \neq 0$), and so we may suppose that $bq < a$. Then $a = bq + r > 0$ for some natural number r. It remains to show that $r < b$. But if $b \leq r$ then

$$bm = b(q + 1) = bq + b \leq bq + r = a,$$

in contradiction to the fact that $m \in X$. Hence $r < b$, as required.

The remaining possibility is that $a < 0$. Then $-a > 0$, and so we may write $-a = bq + r$ by the previous case, where $q \geq 0$ and $0 \leq r < b$. If $r = 0$ then $a = -bq$ and the theorem is satisfied since $-q \leq 0$. Suppose therefore that $r > 0$. Then

$$\begin{aligned} a &= -bq - r \\ &= -bq + b - r - b \\ &= -b(q + 1) + b - r. \end{aligned}$$

Note that $-(q + 1) = -q - 1 < 0$. Moreover $b - r < b$ since $r > 0$, and $b - r > 0$ since $b > r$.

It remains only to prove the uniqueness of q and r in each case. The uniqueness is clear if $a = 0$, for then $q = 0$ so that $r = 0$. Suppose therefore that $a \neq 0$, and that $a = bq_1 + r_1 = bq_2 + r_2$ where q_1, q_2, r_1, r_2 are integers, $0 \leq r_1 < b$ and $0 \leq r_2 < b$. Then

$$\begin{aligned} 0 &= bq_1 + r_1 - bq_2 - r_2 \\ &= b(q_1 - q_2) + r_1 - r_2. \end{aligned}$$

We may suppose without loss of generality that $r_1 \geq r_2$. Therefore $0 \leq r_1 - r_2 \leq r_1 < b$, and so we conclude by the previous case that $q_1 - q_2 = r_1 - r_2 = 0$. Therefore $q_1 = q_2$ and $r_1 = r_2$, as required. □

The integers q and r in Theorem 4.31 are called the **quotient** and **remainder**, respectively, of a on division by b.

If $a = 2q$ for some integer q, then we say that a is **even**. Otherwise a is **odd**. In this case the division theorem shows that $a = 2q + 1$ for some integer q.

Theorem 4.32 *Let $(a, b) \in \mathbb{N} \times \mathbb{N}$, where $b > 1$. Then a can be written uniquely in the form*

$$\sum_{k=0}^{c} a_k b^k,$$

where

(a) *c is a non-negative integer,*
(b) *for each $k \in [0, c]$, a_k is an integer such that $0 \leq a_k < b$ and*
(c) *$a_c > 0$.*

Proof. We prove the existence of the required representation first.

Suppose as an inductive hypothesis that every positive integer less than a can be written in the required form. By the division theorem we can write $a = bq + r$ where q and r are non-negative integers and $r < b$. If $q = 0$ then $a = r$, so that $r > 0$. Since $b^0 = 1$ and $0 < r < b$, we then see that in this case a is of the required form, with $c = 0$ and $a_0 = r$. Suppose therefore that $q > 0$. As $b > 1$, we have $q < bq \leq bq + r = a$. By the inductive hypothesis we can therefore write

$$q = \sum_{k=0}^{c} q_k b^k$$

where c is a non-negative integer, q_k is an integer such that $0 \leq q_k < b$ for each $k \geq 0$ and $q_c > 0$. Thus

$$\begin{aligned} a &= bq + r \\ &= b \sum_{k=0}^{c} q_k b^k + r \\ &= \sum_{k=0}^{c} q_k b^{k+1} + r \\ &= \sum_{k=1}^{c+1} q_{k-1} b^k + r \\ &= \sum_{k=0}^{c+1} a_k b^k \end{aligned}$$

where $a_0 = r$ and $a_k = q_{k-1}$ for each $k > 0$. Since $0 \leq r < b$, $0 \leq a_k < b$ for each $k > 0$ and $a_{c+1} = q_c > 0$, we have written a in the required form.

In order to prove the uniqueness of the expression for a, suppose that

$$a = \sum_{k=0}^{c} a_k b^k = \sum_{k=0}^{d} e_k b^k$$

where c and d are non-negative integers, $0 \leq a_k < b$ and $0 \leq e_k < b$ for each k, $a_c > 0$ and $e_d > 0$. We assume without loss of generality that $c \leq d$.

Suppose first that $c < d$. We may then write

$$a = \sum_{k=0}^{d} a_k b^k$$

where $a_k = 0$ for all $k > c$. In this case it would suffice to show that $a_k = e_k$ for all $k \leq d$, for the fact that $e_d > 0$ would then contradict the equation $a_d = 0$.

It is therefore enough to assume that $c = d$ and show that $a_k = e_k$ for all $k \leq c$, since the case where $c \neq d$ would yield a contradiction as in the previous paragraph. We proceed by induction on c. The requisite result certainly holds if $c = 0$, for then $a_0 = a = e_0$. We may therefore suppose that $c > 0$ and that the result holds with c replaced by $c - 1$.

First we show that $a_0 = e_0$. Suppose without loss of generality that $0 \leq a_0 \leq e_0 < b$, so that $0 \leq e_0 - a_0 < b$. Since

$$\sum_{k=0}^{c} a_k b^k - \sum_{k=0}^{c} e_k b^k = 0$$

we have

$$e_0 - a_0 = \sum_{k=1}^{c} a_k b^k - \sum_{k=1}^{c} e_k b^k$$

$$= b\left(\sum_{k=1}^{c} a_k b^{k-1} - \sum_{k=1}^{c} e_k b^{k-1}\right)$$

$$= bq,$$

where

$$q = \sum_{k=0}^{c-1} a_{k+1} b^k - \sum_{k=0}^{c-1} e_{k+1} b^k.$$

Thus $0 \le bq < b$. It follows that $q = 0$, so that $0 = bq = e_0 - a_0$ and $a_0 = e_0$, as required. Since $q = 0$ we have

$$\sum_{k=0}^{c-1} a_{k+1} b^k = \sum_{k=0}^{c-1} e_{k+1} b^k.$$

It therefore follows from the inductive hypothesis that $a_k = e_k$ for all $k > 0$, and the proof is complete. □

The expression $\sum_{k=0}^{c} a_k b^k$ of Theorem 4.32 is called the **representation** of a with **base** b. The natural number a is often written as $(a_c a_{c-1} \cdots a_0)_b$. For example,

$$(10)_b = 1 \cdot b^1 + 0 \cdot b^0 = b.$$

By convention we usually take $b = (1010)_2$. When this base is used, we write $a_c a_{c-1} \cdots a_0$ instead of $(a_c a_{c-1} \cdots a_0)_b$. Thus $b = 10$ in this notation, which is called the **decimal** notation. For example,

$$263 = 2 \cdot 10^2 + 6 \cdot 10 + 3.$$

Exercises 4

(1) Let $(a, b) \in \mathbb{Z} \times \mathbb{Z}$, where $b \ne 0$. Show that there exist unique integers q and r such that $a = bq + r$ and $0 \le r < |b|$.
(2) Use the telescoping property to prove the following for each $n \in \mathbb{N}$:
   (a) $6 \sum_{k=1}^{n} k^2 = n(n+1)(2n+1)$;
   (b) $4 \sum_{k=1}^{n} k^3 = n^2(n+1)^2$;
   (c) $(a-1) \sum_{k=0}^{n} a^k = a^{n+1} - 1$ for any $a \in \mathbb{Z}$.

(3) Let $a_1, a_2, \cdots, a_n$ be non-negative integers, where $n \in \mathbb{Z}$. Use induction to prove that
$$\prod_{k=1}^{n}(1+a_k) \geq 1 + \sum_{k=1}^{n} a_k$$
for each $n \in \mathbb{N}$. Deduce that if b is a non-negative integer then
$$(1+b)^n \geq 1 + bn.$$

(4) Let n be a natural number greater than 1. If $(a,b) \in \mathbb{Z} \times \mathbb{Z}$, we write $a \equiv b \pmod{n}$ if $a - b = qn$ for some integer q. In this case we say that a is **congruent** to b modulo n. Prove that congruence modulo n is an equivalence relation on $\mathbb{Z}$.

(5) Let a, b, c, d be integers and n an integer greater than 1. If $a \equiv b \pmod{n}$ and $c \equiv d \pmod{n}$, prove that:
  (a) $a + c \equiv b + d \pmod{n}$;
  (b) $a - c \equiv b - d \pmod{n}$;
  (c) $ac \equiv bd \pmod{n}$;
  (d) $a^m \equiv b^m \pmod{n}$ for each $m \in \mathbb{N}$.

(6) Find the decimal representations for the following integers:
  (a) $(101101)_2$;
  (b) $(30212)_4$;
  (c) $(4042)_5$.

(7) Let $(R, +, \cdot, 0, 1, <)$ be an ordered integral domain, and let $(a,b) \in R \times R$. Prove that $|a - b| \geq ||a| - |b||$.

(8) Let $(R, +, \cdot, 0, 1, <)$ be an ordered integral domain, let $n \in \mathbb{N}$ and let $a_k \in R$ for each $k \in [1, n]$. Prove that
$$\left| \prod_{k=1}^{n} a_k \right| = \prod_{k=1}^{n} |a_k|$$
and
$$\left| \sum_{k=1}^{n} a_k \right| \leq \sum_{k=1}^{n} |a_k|.$$

Chapter 5

Rational Numbers

5.1 Definition

Let $(a, b) \in \mathbb{N} \times \mathbb{N}$, where $b > 1$. Then $ab \geq b$, since $a \geq 1$. Thus if c is a positive integer less than b then there is no integer d such that $db = c$, for if $d > 0$ then $db \geq b > c$, if $d = 0$ then $db = 0 < c$, and if $d < 0$ then $db < 0 < c$. This observation motivates the development of yet another kind of number.

As in the case of integers, we begin by defining an equivalence relation. This time, however, the equivalence relation is on $\mathbb{Z} \times (\mathbb{Z} - \{0\})$. Let a, b, c, d be integers such that $b \neq 0$ and $d \neq 0$. In this chapter we define $(a, b) \sim (c, d)$ to mean that $ad = bc$. For example, $(1, 2) \sim (2, 4)$, since $1 \cdot 4 = 2 \cdot 2 = 4$. Note that $(a, b) \sim (ac, bc)$ if $c \neq 0$ and $b \neq 0$. For instance, $(c, c) \sim (1, 1)$.

Theorem 5.1 *The relation $\sim$ is an equivalence relation on $\mathbb{Z} \times (\mathbb{Z} - \{0\})$.*

Proof. It is immediate from the commutativity of multiplication that $(a, b) \sim (a, b)$ for each $(a, b) \in \mathbb{Z} \times (\mathbb{Z} - \{0\})$. Thus reflexivity holds. Symmetry follows similarly from the symmetry of equality and the commutativity of multiplication. Suppose therefore that $(a, b) \sim (c, d)$ and $(c, d) \sim (e, f)$. Thus a, b, c, d, e, f are integers, $b \neq 0$, $d \neq 0$, $f \neq 0$, $ad = bc$ and $cf = de$. Hence $adf = bcf = bde$, and since $d \neq 0$ it follows by the cancellation law (Theorem 4.12) that $af = be$. In other words, $(a, b) \sim (e, f)$. Thus transitivity holds. □

Let $\mathbb{Q}$ be the set of equivalence classes under $\sim$. The elements of $\mathbb{Q}$ are called **rational numbers**. For example, $[(1, 2)] = \{(1, 2), (2, 4), \cdots\}$. We write $0 = [(0, 1)]$ and $1 = [(1, 1)]$. These new uses for the symbols "0" and "1" will be reconciled with the old later. We note that $(a, b) \in 0$ if and only if $a = 0$, for that is the condition under which $(a, b) \sim (0, 1)$. Similarly $(a, b) \in 1$ if and only if $a = b$. For example, $(1, 1) \notin 0$ but $(1, 1) \in 1$, and so we conclude that $0 \neq 1$. The rational number $[(a, b)]$ is usually written as $\frac{a}{b}$ or a/b. Thus $0 = 0/1$ and $1 = 1/1$. Moreover, if $bc \neq 0$ then

$$\frac{ac}{bc} = \frac{a}{b}.$$

For example, $1/2 = 2/4 = (-2)/(-4)$. Note also that $a/a = 1/1 = 1$ if $a \neq 0$.

5.2 Definition of Addition and Multiplication

As in the case of integers, the definitions of addition and multiplication require preliminary lemmas for their justification.

Lemma 5.2 *Let a, b, c, d, e, f, g, h be integers such that $bdfh \neq 0$. If $(a, b) \sim (e, f)$ and $(c, d) \sim (g, h)$, then*

$$(ad + bc, bd) \sim (eh + fg, fh).$$

Proof. Note first that $bd \neq 0$ and $fh \neq 0$.

We are given that $af = be$ and $ch = dg$. Therefore

$$\begin{aligned} fh(ad + bc) &= adfh + bcfh \\ &= bdeh + bdfg \\ &= bd(eh + fg), \end{aligned}$$

as required. □

Let $q = [(a, b)]$ and $r = [(c, d)]$, where a, b, c, d are integers and $bd \neq 0$. We define

$$q + r = [(ad + bc, bd)].$$

By Lemma 5.2, if e, f, g, h are integers such that $fh \neq 0$, $(a, b) \sim (e, f)$ and $(c, d) \sim (g, h)$, then

$$(ad + bc, bd) \sim (eh + fg, fh).$$

Thus $(ad + bc, bd)$ and $(eh + fg, fh)$ determine the same equivalence class $q + r$. In other words, $q + r$ is well defined. We call $q + r$ the **sum** of q and

r, and we refer to the binary operation on $\mathbb{Q}$ denoted by $+$ as **addition**. Note that
$$q + r = \frac{a}{b} + \frac{c}{d} = \frac{ad + bc}{bd}.$$

Lemma 5.3 *Let a, b, c, d, e, f, g, h be integers such that $bdfh \neq 0$. If $(a, b) \sim (e, f)$ and $(c, d) \sim (g, h)$, then*
$$(ac, bd) \sim (eg, fh).$$

Proof. Again $bd \neq 0$, $fh \neq 0$, $af = be$ and $ch = dg$. Therefore $acfh = bdeg$, as required. □

Let $q = [(a, b)]$ and $r = [(c, d)]$, where a, b, c, d are integers such that $bd \neq 0$. We define $q \cdot r = [(ac, bd)]$. Thus Lemma 5.3 shows that $q \cdot r$ is well defined. We call $q \cdot r$ the **product** of q and r, and we refer to the binary operation on $\mathbb{Q}$ denoted by $\cdot$ as **multiplication**. We usually write qr instead of $q \cdot r$. Note that
$$qr = \frac{a}{b} \cdot \frac{c}{d} = \frac{ac}{bd}.$$

Theorem 5.4 $(\mathbb{Q}, +, \cdot, 0, 1)$ *is a commutative ring with unity.*

Proof. Let $(q, r, s) \in \mathbb{Q} \times \mathbb{Q} \times \mathbb{Q}$, $(a, b) \in q$, $(c, d) \in r$ and $(e, f) \in s$. We begin with the following straightforward calculations:

$$\begin{aligned} q + r &= [(ad + bc, bd)] \\ &= [(cb + da, db)] \\ &= r + q; \end{aligned}$$

$$\begin{aligned} (q + r) + s &= [(ad + bc, bd)] + [(e, f)] \\ &= [(f(ad + bc) + bde, bdf)] \\ &= [(fad + fbc + bde, bdf)], \end{aligned}$$

$$\begin{aligned} q + (r + s) &= (r + s) + q \\ &= [(bcf + bde + dfa, dfb)] \\ &= (q + r) + s; \end{aligned}$$

$$\begin{aligned} q + 0 &= [(a, b)] + [(0, 1)] \\ &= [(a, b)] \\ &= q; \end{aligned}$$

$$q + [(-a, b)] = [(a, b)] + [(-a, b)]$$
$$= [(ab - ba, b^2)]$$
$$= [(0, b^2)]$$
$$= 0;$$

$$qr = [(a, b)][(c, d)]$$
$$= [(ac, bd)]$$
$$= [(ca, db)]$$
$$= rq;$$

$$q(rs) = [(a, b)][(ce, df)]$$
$$= [(a(ce), b(df))]$$
$$= [((ac)e, (bd)f)]$$
$$= [(ac, bd)][(e, f)]$$
$$= (qr)s;$$

$$q \cdot 1 = [(a, b)][(1, 1)]$$
$$= [(a, b)]$$
$$= q.$$

Next,
$$q(r + s) = [(a, b)][(cf + de, df)]$$
$$= [(a(cf + de), bdf)]$$
$$= [(acf + ade, bdf)],$$

whereas
$$qr + qs = [(ac, bd)] + [(ae, bf)]$$
$$= [(acbf + bdae, b^2df)].$$

But
$$(acf + ade, bdf) \sim (b(acf + ade), b^2df)$$
$$= (acbf + bdae, b^2df),$$

and so
$$[(acf + ade, bdf)] = [(acbf + bdae, b^2df)],$$

whence
$$q(r+s) = qr + qs.$$
As we have already shown that $0 \neq 1$, the proof is complete. □

The proof above shows that $[(-a,b)]$ is the additive inverse of $[(a,b)]$. Since
$$\begin{aligned}[(a,b)] + [(a,-b)] &= [(-ab+ba, -b^2)] \\ &= [(0,-b^2)] \\ &= 0,\end{aligned}$$
it may also be written as $[(a,-b)]$. We also have the following result.

Theorem 5.5 *If $q \in \mathbb{Q} - \{0\}$ then there exists $r \in \mathbb{Q}$ such that $qr = 1$.*

Proof. Choose $(a,b) \in q$. Then $a \neq 0$, since $q \neq 0$, and so
$$(b,a) \in \mathbb{Z} \times (\mathbb{Z} - \{0\}).$$
Moreover
$$[(a,b)][(b,a)] = [(ab,ba)] = 1.$$
Consequently the theorem holds with $r = [(b,a)]$. □

Note that if $(a,b) \in \mathbb{Z} \times \mathbb{Z}$ then
$$[(a,1)] + [(b,1)] = [(a+b,1)]$$
and
$$[(a,1)][(b,1)] = [(ab,1)].$$
Let us define $\Phi(a) = [(a,1)]$ for each $a \in \mathbb{Z}$. The foregoing calculations show that $\Phi(a) + \Phi(b) = \Phi(a+b)$ and $\Phi(a)\Phi(b) = \Phi(ab)$. We also see that Φ is an injection from $\mathbb{Z}$ into $\mathbb{Q}$. Indeed, suppose that $\Phi(a) = \Phi(b)$. Then $[(a,1)] = [(b,1)]$. Hence $(a,1) \sim (b,1)$, and so $a = b$ by definition.

Previously we saw how the natural numbers can be incorporated amongst the integers. Similarly we now see that rational numbers of the form $[(a,1)]$, where $a \in \mathbb{Z}$, can be identified with integers. Moreover $\Phi(0) = [(0,1)] = 0$ and $\Phi(1) = [(1,1)] = 1$. Thus 0 and 1 in $\mathbb{Q}$ are identified with 0 and 1, respectively, in $\mathbb{Z}$. Since
$$[(a,1)][(1,b)] = [(a,b)]$$

whenever $(a, b) \in \mathbb{Z} \times (\mathbb{Z} - \{0\})$, it follows that $a(1/b) = a/b$ whenever $b \neq 0$. In particular $a(1/a) = 1$ whenever $a \neq 0$.

5.3 Fields

We are now ready to generalize our results, as we did previously in the case of integers. Let $(F, +, \cdot, 0, 1)$ be a commutative ring with unity. Suppose that for each $q \in F - \{0\}$ there exists $r \in F$ such that $qr = 1$. Then $(F, +, \cdot, 0, 1)$ is called a **field**. For example, $(\mathbb{Q}, +, \cdot, 0, 1)$ is a field.

If $(F, +, \cdot, 0, 1)$ is a field and $q \in F - \{0\}$ let us show that there exists a unique $r \in F$ such that $qr = 1$. The existence of r is guaranteed by definition. Suppose that $qr = qs = 1$, where $(r, s) \in F \times F$. Then $r = 1 \cdot r = qsr = qrs = 1 \cdot s = s$, as required. We call r the **multiplicative inverse** of q, and denote it by q^{-1}. Thus $q \cdot q^{-1} = 1$. As an example, since $1 \cdot 1 = 1$ we have $1^{-1} = 1$.

If $(F, +, \cdot, 0, 1)$ is a field, $q \in F$ and $r \in F - \{0\}$, then we define

$$\frac{q}{r} = qr^{-1}.$$

This element of F is called the **quotient** of q by r, and the operation denoted by $/$ is termed **division**. Observe that $r^{-1} = 1 \cdot r^{-1} = 1/r$, $q/1 = q \cdot 1^{-1} = q \cdot 1 = q$, $0/r = 0 \cdot r^{-1} = 0$, $r/r = r \cdot r^{-1} = 1$ and $q/r = qr^{-1} = q(1/r)$.

We now turn to the general properties of fields.

Theorem 5.6 *Every field is an integral domain.*

Proof. Let $(F, +, \cdot, 0, 1)$ be a field. Thus $(F, +, \cdot, 0, 1)$ is a commutative ring with unity. Let q and r be nonzero elements of F. It suffices to prove that $qr \neq 0$. Suppose therefore that $qr = 0$. Then $q^{-1}qr = 0$, and since $q^{-1}q = 1$ we obtain the contradiction that $r = 0$. □

Thus all the properties enjoyed by integral domains carry over to fields. The additional properties that need to be mentioned concern multiplicative inverses.

Let $(F, +, \cdot, 0, 1)$ be a field, and let $r \in F - \{0\}$. Since $r \cdot r^{-1} = 1 \neq 0$, it follows that $1/r \neq 0$. We also deduce from the uniqueness of r^{-1} that $1/(1/r) = (r^{-1})^{-1} = r$. If $t \in F - \{0\}$, then $rtr^{-1}t^{-1} = rr^{-1}tt^{-1} = 1$, and so $1/rt = (1/r)(1/t)$ by the uniqueness of the multiplicative inverse. More generally, if $(q, s) \in F \times F$ then $(q/r)(s/t) = qr^{-1}st^{-1} = qs(rt)^{-1} = qs/rt$.

Thus $(r/t)(t/r) = rt/rt = 1$, and so $1/(r/t) = (r/t)^{-1} = t/r$. Furthermore, if $q/r = s/t$, then multiplication of both sides by rt yields $qt = rs$. (This procedure is sometimes called cross multiplication.) Conversely if $qt = rs$ then multiplication of both sides by $1/rt$ (in other words, division by rt) gives $q/r = s/t$. Thus if $q = rs$ then $s = q/r$. Moreover $q/r = qt/rt$, since $qrt = rqt$. Hence

$$\frac{q}{r} + \frac{s}{t} = \frac{qt}{rt} + \frac{rs}{rt}$$

$$= qt(rt)^{-1} + rs(rt)^{-1}$$

$$= (qt + rs)(rt)^{-1}$$

$$= \frac{qt + rs}{rt}.$$

Also, if $s \neq 0$ then $(q/r)/(s/t) = (q/r)(1/(s/t)) = (q/r)(t/s) = qt/rs$.

5.4 Exponentiation

The concept of exponentiation discussed in the previous chapter can now be extended somewhat. Let $(F, +, \cdot, 0, 1)$ be a field, and let $q \in F - \{0\}$. For any non-negative integer a, we define $q^a = \prod_{k=1}^{a} q$, in agreement with our previous definition. Thus $q^1 = q$ and $q^0 = 1$. We also defined q^{-1} in the previous section. Recalling that $q^{-1} = 1/q$, we define $q^{-a} = 1/q^a$. Thus $q^a q^{-a} = 1$, and we also have both $q^a = 1/q^{-a}$ and $q^{-a} = 1/q^a$. It follows from the last equation that if a is a negative integer then $q^a = 1/q^{-a}$. This equation therefore holds for any integer a. We now show that the algebraic rules for exponentiation carry over in this setting.

Theorem 5.7 *Let $(F, +, \cdot, 0, 1)$ be a field. Let $q \in F - \{0\}$, $r \in F - \{0\}$ and $a \in \mathbb{Z}$. Then*

$$(qr)^a = q^a r^a.$$

Proof. This result has already been established for each $a \geq 0$ in Theorem 4.13. If $a < 0$, then $-a > 0$, and so

$$(qr)^a = \frac{1}{(qr)^{-a}} = \frac{1}{q^{-a}r^{-a}} = \frac{1}{q^{-a}} \cdot \frac{1}{r^{-a}} = q^a r^a.$$

□

Theorem 5.8 *Let $(F, +, \cdot, 0, 1)$ be a field, let $q \in F - \{0\}$ and let $(a, b) \in \mathbb{Z} \times \mathbb{Z}$. Then*

$$q^a q^b = q^{a+b}.$$

Proof. This result has already been established if $a \geq 0$ and $b \geq 0$ in Theorem 4.14 (a).

If $a < 0$ and $b < 0$, then $-a > 0$, $-b > 0$ and $a + b < 0$. In this case

$$q^a q^b = \frac{1}{q^{-a}} \cdot \frac{1}{q^{-b}} = \frac{1}{q^{-a}q^{-b}} = \frac{1}{q^{-a-b}} = \frac{1}{q^{-(a+b)}} = q^{a+b}.$$

We may now assume without loss of generality that $a \geq 0$ and $b < 0$. Thus $-b > 0$. If $a + b \geq 0$, then $q^{a+b}q^{-b} = q^{a+b-b} = q^a$, and the desired result follows upon multiplication of both sides by q^b. If $a + b < 0$, then $q^{-(a+b)}q^a = q^{-a-b+a} = q^{-b}$, and multiplication of both sides by $q^{a+b}q^b$ again yields $q^a q^b = q^{a+b}$. □

Thus

$$\frac{q^a}{q^b} = q^a q^{-b} = q^{a-b}.$$

Theorem 5.9 *Let $(F, +, \cdot, 0, 1)$ be a field, let $q \in F - \{0\}$ and let $(a, b) \in \mathbb{Z} \times \mathbb{Z}$. Then*

$$(q^a)^b = q^{ab}.$$

Proof. If $b \geq 0$, we can use induction on b. Observe first that both sides are equal to 1 if $b = 0$. Suppose as an inductive hypothesis that $(q^a)^b = q^{ab}$ for some non-negative integer b. Then

$$\begin{aligned}(q^a)^{b+1} &= (q^a)^b q^a \quad &\text{(by Theorem 5.8)} \\ &= q^{ab} q^a \\ &= q^{ab+a} \\ &= q^{a(b+1)},\end{aligned}$$

as required.

If $b < 0$, then $-b > 0$, and so

$$(q^a)^b = \frac{1}{(q^a)^{-b}} = \frac{1}{q^{-ab}} = q^{ab},$$

as required. □

Exponentiation

We now derive a formula for $(a+b)^n$ where a and b are members of a field F and n is a non-negative integer. First let us define

$$\binom{n}{r} = \frac{n!}{r!(n-r)!},$$

where n and r are non-negative integers and $r \leq n$. For example,

$$\binom{n}{0} = \frac{n!}{0!n!} = 1.$$

Similarly

$$\binom{n}{n} = 1.$$

We obtain the following lemma.

Lemma 5.10 *If n and r are positive integers and $r \leq n$ then*

$$\binom{n+1}{r} = \binom{n}{r} + \binom{n}{r-1}.$$

Proof. By direct calculation we have

$$\binom{n}{r} + \binom{n}{r-1} = \frac{n!}{r!(n-r)!} + \frac{n!}{(r-1)!(n-r+1)!}$$
$$= \frac{(n-r+1)n! + rn!}{r!(n-r+1)!}$$
$$= \frac{(n+1)n!}{r!(n-r+1)!}$$
$$= \frac{(n+1)!}{r!(n-r+1)!}$$
$$= \binom{n+1}{r}.$$

$\square$

The following result is known as the binomial theorem.

Theorem 5.11 **(Binomial Theorem)** *Let F be a field, let $(a,b) \in F \times F$ and let n be a non-negative integer. Then*

$$(a+b)^n = \sum_{k=0}^{n} \binom{n}{k} a^k b^{n-k},$$

where the convention that $0^0 = 1$ is used.

Proof. Both sides are equal to 1 if $n = 0$. Assume that the theorem holds for some integer $n \geq 0$. Then

$$(a+b)^{n+1} = (a+b)(a+b)^n$$
$$= (a+b)\sum_{k=0}^{n}\binom{n}{k}a^k b^{n-k}$$
$$= \sum_{k=0}^{n}\binom{n}{k}a^{k+1}b^{n-k} + \sum_{k=0}^{n}\binom{n}{k}a^k b^{n-k+1}$$
$$= \sum_{k=1}^{n+1}\binom{n}{k-1}a^k b^{n-k+1} + \sum_{k=0}^{n}\binom{n}{k}a^k b^{n-k+1}$$
$$= \sum_{k=1}^{n}\binom{n}{k-1}a^k b^{n+1-k} + a^{n+1} + b^{n+1} + \sum_{k=1}^{n}\binom{n}{k}a^k b^{n+1-k}$$
$$= b^{n+1} + \sum_{k=1}^{n}\left(\binom{n}{k-1} + \binom{n}{k}\right)a^k b^{n+1-k} + a^{n+1}$$
$$= b^{n+1} + \sum_{k=1}^{n}\binom{n+1}{k}a^k b^{n+1-k} + a^{n+1}$$
$$= \sum_{k=0}^{n+1}\binom{n+1}{k}a^k b^{n+1-k}.$$

The result follows by induction. □

5.5 Order

We now generalize the concept of order discussed previously for the case of integers. We begin with a lemma.

Lemma 5.12 *Let a, b, c, d be integers such that $bd \neq 0$ and $(a,b) \sim (c,d)$. Then $ab > 0$ if and only if $cd > 0$.*

Proof. Suppose $ab > 0$. Since $(a,b) \sim (c,d)$, we find that $ad = bc$. Hence $abd^2 = b^2 cd$. Since $bd \neq 0$ we have $b^2 > 0$ and $d^2 > 0$, and as $ab > 0$ by hypothesis we deduce that $cd > 0$. Similarly if $cd > 0$ then $ab > 0$. □

By replacing a and c by $-a$ and $-c$ respectively, we find that $ab < 0$ if and only if $cd < 0$, for if $(a,b) \sim (c,d)$ then $(-a,b) \sim (-c,d)$.

Let q be a rational number. Then $q = a/b$ for some integers a and b such that $b \neq 0$. If $ab = 0$, then $a = 0$, so that $q = 0$. Conversely if $q = 0$ then $ab = 0$. We say that q is **positive** if $ab > 0$ and **negative** if $ab < 0$. Lemma 5.12 guarantees that these concepts are well defined. As in the case of integers, 0 is neither positive nor negative, but any nonzero rational number is either positive or negative, though not both. Moreover q is negative if and only if $-q$ is positive. If $q = a$, then $b = 1$, and so q is positive if $a > 0$ and negative if $a < 0$. Thus our definitions of positive and negative rational numbers agree with our earlier definitions of positive and negative integers. In any case we write $q > 0$ if q is positive and $q < 0$ if q is negative. If $q \neq 0$ then we say that the **sign** of q is positive or negative according to whether $q > 0$ or $q < 0$. If q and r are rational numbers, then we say that q is **less** than r, or that r is **greater** than q, if $r - q > 0$. In this case we write $q < r$ or $r > q$. Note that r is positive if and only if r is greater than 0, in agreement with our previous notation. Similarly q is negative if and only if it is less than 0. These definitions are also in agreement with our earlier definitions in the case of integers. As in the case of integers we write $q \leq r$ or $r \geq q$ if either $q < r$ or $q = r$.

Now let q and r be positive rational numbers. Suppose that $q = a/b$ and $r = c/d$, where a, b, c, d are integers and $bd \neq 0$. Then

$$q + r = \frac{a}{b} + \frac{c}{d} = \frac{ad + bc}{bd}.$$

Now $ab > 0$ and $cd > 0$, since $q > 0$ and $r > 0$, and so

$$bd(ad + bc) = abd^2 + b^2cd > 0,$$

because $b^2 > 0$ and $d^2 > 0$ since $b \neq 0$ and $d \neq 0$. Therefore $q + r > 0$. Similarly $qr = ac/bd > 0$ since $abcd > 0$.

We have now taken the theory to the point where, as in the case of integers, we can verify that $(\mathbb{Q}, +, \cdot, 0, 1, <)$ is an ordered integral domain. The axioms for an ordered integral domain are checked exactly as in the case of integers, and the verification is therefore omitted.

An ordered integral domain $(F, +, \cdot, 0, 1, <)$ is called an **ordered field** if $(F, +, \cdot, 0, 1)$ is a field. Thus $(\mathbb{Q}, +, \cdot, 0, 1, <)$ is an ordered field. Note that ordered fields must satisfy all the properties enjoyed by ordered integral domains.

Let $(F, +, \cdot, 0, 1, <)$ be an ordered field and choose $q \in F - \{0\}$. Then $qq^{-1} = 1 > 0$, and so $1/q \neq 0$, and q and $1/q$ have the same sign. Now suppose that $0 < q < r$. Since r and $1/r$ must have the same sign, we have

$1/r > 0$. Similarly $1/q > 0$, and so $1/qr > 0$. Multiplication of both sides of the inequality $q < r$ by $1/qr$ thus gives $1/r < 1/q$.

A similar argument shows that if $q < r < 0$ then again $1/r < 1/q$. On the other hand, if $q < 0 < r$ then $1/q < 0 < 1/r$.

Now choose $q \in F$ and $r \in F - \{0\}$. Then $q = qr/r$, and so

$$|q| = \left|\frac{q}{r}\right| |r|.$$

Since $r \neq 0$, we have $|r| \neq 0$. Division by $|r|$ therefore gives

$$\left|\frac{q}{r}\right| = \frac{|q|}{|r|}.$$

We prove one final property of ordered fields, which is referred to as **density**. We define $2 = 1 + 1$ in any ordered field.

Theorem 5.13 *Let $(F, +, \cdot, 0, 1, <)$ be an ordered field, and let q and r be elements of F such that $q < r$. Then there exists $s \in F$ such that $q < s < r$.*

Proof. Since $q < r$ we have $2q < r + q < 2r$. Division by 2 (that is, multiplication by $1/2$) yields

$$q < \frac{r+q}{2} < r,$$

and so the desired result holds with $s = (r+q)/2$. (Observe that if $1+1 = 0$, then we have the contradiction that $1 = 0 + 1 < 1 + 1 = 0$.) □

Roughly speaking, what this result says is that between any two distinct elements of an ordered field there lies another. Of course it is not unique. For example we have $\frac{1}{2} > \frac{1}{3} > \frac{1}{5}$ and $\frac{1}{2} > \frac{1}{4} > \frac{1}{5}$.

We conclude this section with some examples in which we are required to solve a given inequality for a variable x, which is assumed to be a rational number.

Example 5.1 Solve

$$2x - 1 \leq 4x + 4.$$

Solution: Adding $-4 - 2x$ to both sides gives $2x \geq -5$, whence

$$x \geq -\frac{5}{2}.$$

△

Example 5.2 Solve

$$x^2 < 3x - 2.$$

Solution: Adding $2 - 3x$ to both sides gives $x^2 - 3x + 2 < 0$. Therefore $(x - 2)(x - 1) < 0$. Exactly one of $x - 1$ and $x - 2$ must be negative, and since $x - 2 < x - 1$ it follows that $x - 2 < 0 < x - 1$. Hence $1 < x < 2$.

△

Example 5.3 Solve

$$\frac{x-2}{x+1} \geq 0.$$

Solution: Note that we must have $x + 1 \neq 0$. It follows that either $x - 2 \geq 0$ and $x + 1 > 0$, or $x - 2 \leq 0$ and $x + 1 < 0$. The former case holds if and only if $x \geq 2$, the latter if and only if $x < -1$.

△

5.6 The Archimedean Property and Certain Technical Results

This section is devoted primarily to proving a technical result that will be of importance in Section 6.4. The results in this section are of interest in their own right and provide further insight into the rational numbers. Our choice of material is motivated by the results needed to prove Theorem 5.18, which will be needed in Section 6.4. Eventually, the chain of logic will illustrate the inadequacy of the rational numbers and pave the way for the construction of the real numbers.

If r is a positive rational number and s is any rational number, then it is intuitively obvious that for all $n \in \mathbb{N}$ "sufficiently large" we have $s/n < r$. Formally, we have the following result.

Theorem 5.14 (**Archimedean Property**) *Let r and s be rational numbers. If $r > 0$ then there is an $N \in \mathbb{N}$ such that $s/n < r$ for all $n > N$.*

Proof. If $s \leq 0$, then the result holds with $N = 1$. Suppose now that $s > 0$. Since r and s are positive rational numbers there are positive integers a, b, c, d such that $r = a/b$ and $s = c/d$. Choose $N = c(b+1)$. Then $1/n < 1/N$ for all $n > N$ and consequently

$$\frac{s}{n} < \frac{s}{N} = \frac{c}{dc(b+1)} = \frac{1}{d(b+1)}$$

whenever $n > N$. Now $d \geq 1$, so that $1/d \leq 1$, and $b + 1 > b$ so that $1/(b+1) < 1/b$. Therefore $s/n < 1/b$ for all $n > N$, and since $a \geq 1$ we have

$$\frac{s}{n} < \frac{1}{b} \leq \frac{a}{b} = r$$

whenever $n > N$. □

The case where $r = 1$ yields the following corollary.

Corollary 5.15 *For any $s \in \mathbb{Q}$ there exists a positive integer $n > s$.*

The next theorem refines the density result given in Theorem 5.13.

Theorem 5.16 *Let r and s be any rational numbers such that $r < s$. For any positive rational number p there exist an $m \in \mathbb{N}$ and $m - 2$ rational numbers $q_2, q_3, \ldots, q_{m-1}$ such that, for $q_1 = r$ and $q_m = s$,*

$$0 < q_k - q_{k-1} < p$$

for each $k \in \{2, 3, \ldots, m\}$.

Proof. Choose any positive rational number p. Now $s - r \in \mathbb{Q}$ and Theorem 5.14 implies that there is an $N \in \mathbb{N}$ such that

$$\frac{s-r}{m} < p$$

whenever $m > N$. For each integer $m > N + 1$ it follows that

$$\frac{s-r}{m-1} < p.$$

Let

$$q_k = q_{k-1} + \frac{s-r}{m-1} \tag{5.1}$$

for each $k \in \{2, 3, \ldots, m-1\}$. Since $q_1 = r$ it follows by induction that

$$q_k = r + \frac{(k-1)(s-r)}{m-1}$$

for each $k \in \{1, 2, \ldots, m-1\}$. Thus

$$\begin{aligned} q_m - q_{m-1} &= s - r - \frac{(m-2)(s-r)}{m-1} \\ &= \frac{(m-1)(s-r) - (m-2)(s-r)}{m-1} \\ &= \frac{s-r}{m-1}, \end{aligned}$$

and we deduce that 5.1 also holds with $k = m$. The required results now follow from the facts that

$$q_k - q_{k-1} = \frac{s-r}{m-1}$$

for all $k \in \{2, 3, \ldots, m\}$ and

$$0 < \frac{s-r}{m-1} < p.$$

□

Lemma 5.17 *Let r be any positive rational number. Then there exist positive rational numbers a and b such that $a^2 < r < b^2$.*

Proof. If $r < 1$ then $r^2 < r < 1$, so that we can choose $a = r$ and $b = 1$. If $r = 1$ then $(1/2)^2 < 1 < 2^2$, so that we can choose $a = 1/2$ and $b = 2$. If $r > 1$ then $1 < r < r^2$, so that we can choose $a = 1$ and $b = r$. □

Theorem 5.18 *Let r be any positive rational number and n any natural number. Then there exists a positive rational number s such that*

$$r < s^2 < r + \frac{1}{n}.$$

Proof. Choose any positive rational number r and any natural number n. Lemma 5.17 implies that there exist positive rational numbers a, b such that $a^2 < r < b^2$. Thus $a < b$, and Theorem 5.16 shows that for any positive rational number p there exist an $m \in \mathbb{N}$ and rational numbers $q_1, q_2, \ldots, q_m$ such that

$$a = q_1 < q_2 < \ldots < q_{m-1} < q_m = b$$

and

$$q_k - q_{k-1} < p$$

for each $k \in \{2, 3, \ldots, m\}$. Since $q_k < q_{k-1} + p$ we have

$$q_k^2 < q_{k-1}^2 + 2q_{k-1}p + p^2,$$

and consequently

$$q_k^2 - q_{k-1}^2 < 2q_{k-1}p + p^2 = p(2q_{k-1} + p)$$

for each $k \in \{2, 3, \ldots, m\}$. Choose

$$p = \frac{1}{n(2b+1)}.$$

Now, $q_{k-1} < q_m = b$ for each $k \in \{2, 3, \ldots, m\}$, and since $b > 0$ we also have $0 < p < 1$. Therefore

$$q_k^2 - q_{k-1}^2 < p(2q_{k-1} + p) < p(2b+1) = \frac{1}{n} \qquad (5.2)$$

for each $k \in \{2, 3, \ldots, m\}$.

Let $M \subset \mathbb{N}$ be the set

$$\{k \in \{1, 2, \ldots, m\} : q_k^2 > r\}.$$

The set M is nonempty because $q_m^2 = b^2 > r$; hence, by Theorem 3.30, M has a least element j. Note that $j > 1$ since $q_1^2 = a^2 < r$. Moreover, $q_j^2 > r$ and $q_{j-1}^2 \leq r$.

We now show that we can choose $s = q_j$. It suffices to verify that $q_j^2 < r + 1/n$. Suppose that $q_j^2 \geq r + 1/n$. Since $r \geq q_{j-1}^2$ we have

$$q_j^2 - q_{j-1}^2 \geq \left(r + \frac{1}{n}\right) - r = \frac{1}{n},$$

and this inequality contradicts inequality 5.2. Therefore $q_j^2 < r + 1/n$, as required. □

Exercises 5

(1) Prove that every finite integral domain is a field.
(2) Let $X = \{0, 1\}$ where $0 \neq 1$, and define $0 + 0 = 1 + 1 = 0$, $0 + 1 = 1 + 0 = 1$, $0 \cdot 0 = 1 \cdot 0 = 0 \cdot 1 = 0$ and $1 \cdot 1 = 1$. Show that $(X, +, \cdot, 0, 1)$ is a field.
(3) Solve the following inequalities for x:
   (a) $4x - 1 > 2x + 3$;
   (b) $4 - 5x \leq 3x - 1$;

(c) $6x + 7 < 3x - 9$;
(d) $8x + 11 \geq 6x + 4$;
(e) $x^2 < 5x - 6$;
(f) $x^2 + 4 > 4x$;
(g) $x \leq x^2 - 6$;
(h) $\dfrac{2x+3}{x+4} < 0$;
(i) $\dfrac{x-1}{2-3x} \leq 0$;
(j) $\dfrac{2-x}{1-3x} \geq 0$.

(4) For any nonzero rational number q, prove that $q + 1/q \geq 2$.

Chapter 6

Sequences on $\mathbb{Q}$

The extension of the rational number system to the real number system involves certain concepts and techniques fundamentally different from those used so far to extend number systems. In particular, it is necessary to introduce sequences and the notion of a limit. We postpone the construction of the real number system until the next chapter. In this chapter we consider sequences of rational numbers.

6.1 Sequences and Limits

A **sequence** on a set X is a function $f : \mathbb{N} \to X$. The elements of the range of the sequence are called the **terms** of the sequence. Some examples of sequences $f : \mathbb{N} \to X$ are:

(1) $f(n) = n^2$, $X = \mathbb{N}$;
(2) $f(n) = (-1)^n n$, $X = \mathbb{Z}$;
(3) $f(n) = 1/n$, $X = \mathbb{Q}$.

Note that the nature of the set X is quite general; X need not be a set consisting of numbers. For example, X can be the set of all circles in the plane with centre at some fixed point. A sequence can be defined by identifying with each $n \in \mathbb{N}$ the circle in X with radius n. For the present purposes, however, X will usually be a set of numbers.

A sequence is a function and as such it can be defined in a number of ways. The function $f : \mathbb{N} \to X$ can be given explicitly as in the above examples or it may be given implicitly by requiring the terms to satisfy some property or relation. For example, the relations $x_1 = 1$, $x_2 = 1$, and $x_{n+1} = x_n x_{n-1} + x_n^2$ determine a sequence on $\mathbb{N}$ but no formula for f is given. Another common way to define a sequence is simply to provide a list $w_1, w_2, \ldots, w_k$ of images of the first k elements in the domain of the sequence. This is generally given in the form $w_1, w_2, \ldots, w_k, \ldots$, where the ellipsis at the end of the list indicates that the process continues *ad infinitum*. For example, $1, 4, 3, 16, 5, 36, \ldots$ defines a sequence ($f(n) = n$ if n is odd; $f(n) = n^2$ if n is even). This method of defining a sequence has obvious pitfalls, but it is sometimes convenient.

The notion of a sequence can be generalized slightly to include functions $f : \mathcal{D}_f \to X$ such that the domain $\mathcal{D}_f$ of f is the set of all integers greater than some fixed integer N. Let $t_N : \mathbb{N} \to \mathbb{Z}$ be the sequence defined by $t_N(n) = N + n$. Clearly $f \circ t_N$ is a sequence and in this sense f may be regarded as a sequence. (Sequences could have been defined in terms of sets such as $\mathcal{D}_f$, but the extra structure present is superfluous.) The above considerations provide flexibility in choosing the minimum element in the sequence domain. In other words, the sequence can be "started" at any integer by reindexing if necessary. Usually the minimum domain element for a sequence is 0 or 1. It will be assumed throughout this chapter that the domain of every sequence under consideration is $\mathbb{N}$.

Sequences can be represented in many ways. Formally, a sequence $f : \mathbb{N} \to X$ can be represented as a set

$$\{(1, f(1)), (2, f(2)), \ldots\}$$

of ordered pairs but this soon becomes cumbersome. To streamline the presentation, the notation $(f(n))$ will be used. The sequence $f : \mathbb{N} \to \mathbb{Q}$ defined by $f(n) = 1/n$, for example, is denoted by $(1/n)$.

It is the behaviour of sequences for arbitrarily large n that is of primary importance in analysis. Suppose f is some sequence and A is some property (such as being a positive integer) that $f(n)$ may or may not have. There are three possibilities:

(1) $f(n)$ has property A for all values of n except perhaps for a finite number of values;
(2) $f(n)$ has property A for at most a finite number of values;
(3) neither (1) nor (2) is true.

For example if A is the property of being a positive integer, then a sequence such as $(-50 + n)$ is of type (1) because $-50 + n > 0$ for all $n > 50$ so that there are only a finite number of terms that do not have this property. On the other hand, a sequence such as $(50 - n)$ is of type (2). A sequence such as $((-1)^n)$ is of type (3).

If a sequence $f : \mathbb{N} \to X$ is of type (1) with respect to some property A, then there can be at most a finite number k of exceptional values of n for which $f(n)$ does not have property A. These exceptional values need not of course correspond to the first k numbers in the domain (though this is frequently the case in practice). What is important is that there are only a finite number of such values and therefore there is some number $N \in \mathbb{N}$ such that $f(n)$ has property A for all values of n greater than N. The value of N evidently depends on the property A, and there may be many values of n less than N for which $f(n)$ has property A. Suppose for example a term has property A if it is a positive integer, and consider the sequence $(n^2 - 11n + 28)$. The terms of the sequence have this property for $n = 1, 2, 3$ and for any value of n greater than 7. Here, a natural choice for N is 7, but any natural number $m > 7$ is also suitable. Though it is sometimes of practical interest to determine the minimum N, we are usually not concerned with this value. What is significant is whether any such number exists.

It is the properties of a sequence that persist for all values of n beyond some N that are of interest in analysis. The investigation can be carried to the natural extreme of examining those properties which can be described only in terms of n growing without bound. Consider, for example, the sequence $(1/n)$. It is plain that given any positive number $q \in \mathbb{Q}$ one can find an N so large that $1/n < q$ for all values of n greater than N. This property can be described loosely by the phrase "as n increases, $1/n$ decreases towards zero". Sequence properties couched in terms of n growing without bound are of primary importance in analysis and will dominate virtually all of this and the next chapter. It is thus desirable to introduce certain phrases and definitions concerning the growth of values of n. In particular, to describe the boundless increase of n the phrase "as n tends towards infinity" will be used. Symbolically, this is denoted by $n \to \infty$, where the symbol ∞ denotes "infinity". The reader is warned that the symbol ∞ as used here does not correspond to a number of any type and carries meaning only in a context which has been defined.

It may be that as $n \to \infty$ a sequence tends towards some number. For example, the sequence $(1/n)$ tends towards zero in the sense that $|1/n - 0|$ can be made arbitrarily small. Similarly, the sequence $(7 - 1/n^2)$ tends towards 7 as $n \to \infty$. In contrast, a sequence such as $((-1)^n n)$ does not tend towards any particular value as $n \to \infty$. The idea of whether a sequence tends towards any particular value as $n \to \infty$ is captured mathematically in the concept of convergence.

Let (a_n) be a sequence on $\mathbb{Q}$ and $a \in \mathbb{Q}$. The sequence (a_n) is said to **converge** to a if for every positive rational number ϵ there exists a natural number N such that $|a_n - a| < \epsilon$ for all values of n greater than N. If (a_n) converges to a, the number a is called the **limit** of the sequence, and this relationship is denoted by $\lim_{n \to \infty} a_n = a$ or simply $\lim a_n = a$. The alternative notation $a_n \to a$ as $n \to \infty$ (or simply $a_n \to a$) will also be used. We say that a_n **approaches** a as $n \to \infty$. If a sequence does not converge it is said to **diverge**. A sequence (a_n) on $\mathbb{Q}$ is **convergent** in $\mathbb{Q}$ if there exists $a \in \mathbb{Q}$ to which (a_n) converges; otherwise (a_n) is **divergent** in $\mathbb{Q}$.

Example 6.1 If the sequence $(1/n)$ converges, a conspicuous candidate for a limit is zero. In order to establish that $\lim_{n \to \infty} 1/n = 0$ it must be shown that for any choice of $\epsilon > 0$ an N can be found such that $|1/n - 0| < \epsilon$ whenever $n > N$.

Choose any $\epsilon > 0$. Let N be any natural number such that $N \geq 1/\epsilon$. Now choose any $n > N$. Then the condition $1/n < 1/N \leq \epsilon$ is satisfied; consequently, $\lim_{n \to \infty} 1/n = 0$.

$\triangle$

Example 6.2 Consider the sequence $(4 + (-1)^n/n^2)$. If this sequence converges, a candidate for a limit is 4. In order to prove that this sequence converges to 4 we need to show that for any choice of $\epsilon > 0$ there exists an N such that $|4 + (-1)^n/n^2 - 4| < \epsilon$ for all $n > N$.

Choose $\epsilon > 0$. For all $n \in \mathbb{N}$, we have $|4 + (-1)^n/n^2 - 4| = 1/n^2$. Let N be any natural number such that $N \geq 1/\epsilon$. If $n > N$ then

$$\left|4 + \frac{(-1)^n}{n^2} - 4\right| = \frac{1}{n^2} < \frac{1}{N^2} \leq \frac{1}{N} \leq \epsilon$$

and therefore

$$\lim_{n \to \infty} \left(4 + \frac{(-1)^n}{n^2}\right) = 4.$$

$\triangle$

Example 6.3 The sequence $((-1)^n)$ diverges. This is somewhat obvious since the terms in this sequence alternate between 1 and -1. To prove that this sequence diverges it must be shown that it cannot converge, and this result can be established by contradiction.

Suppose the sequence converges to some $a \in \mathbb{Q}$. Then for any $\epsilon > 0$ there is an N_1 such that $|(-1)^n - a| < \epsilon$ for all $n > N_1$. In particular, such an N_1 must exist for the choice $\epsilon = 1/2$. Since any $N_2 \geq N_1$ must also have the property that $|(-1)^n - a| < 1/2$ whenever $n > N_2$, it can be assumed without loss of generality that N_1 is even. Choosing $n > N_1$ such that n is even, we find that $|1 - a| < 1/2$, so that $1/2 < a < 3/2$. But $N_1 + 1 > N_1$ and $|(-1)^{N_1+1} - a| = |1 + a| > 1/2$ since a is positive. This inequality contradicts the definition of N_1 and therefore the sequence must diverge.

△

Example 6.4 Let $(a_n) = ((2n+1)/(n+4))$. The terms a_n can be written in the form $a_n = (2 + 1/n)/(1 + 4/n)$, and if n is large then $1/n$ and $4/n$ are small. This observation suggests that the sequence converges to 2.

Choose any $\epsilon > 0$. Note that

$$|a_n - 2| = \left|\frac{2n+1}{n+4} - 2\right| = \frac{7}{n+4}.$$

We must find N such that $7/(n+4) < \epsilon$ for all $n > N$. The required inequality holds if and only if $7 < n\epsilon + 4\epsilon$, and so we may choose any $N \geq (7 - 4\epsilon)/\epsilon$.

△

If a sequence converges to some limit, the question arises as to whether this is the only limit. In other words, is the limit of a convergent sequence unique? The next theorem confirms that such is the case.

Theorem 6.1 *If a sequence on $\mathbb{Q}$ converges, then the limit of the sequence is unique.*

Proof. Let (a_n) be a sequence on $\mathbb{Q}$, and suppose that $\lim_{n \to \infty} a_n = a$ and $\lim_{n \to \infty} a_n = b$. By definition, for any choice of $\epsilon > 0$ there exists an N_1 such that $n > N_1$ implies that $|a_n - a| < \epsilon$. Similarly, there exists an N_2 such that $n > N_2$ implies that $|a_n - b| < \epsilon$. Now

$$|a - b| = |a - a_n + a_n - b|$$
$$\leq |a - a_n| + |a_n - b|,$$

by the triangle inequality. Let $N = \max\{N_1, N_2\}$ and choose $n > N$. Then $n > N_1$ and $n > N_2$, so that $|a_n - a| < \epsilon$ and $|a_n - b| < \epsilon$, and therefore

$$|a - b| < 2\epsilon \qquad (6.1)$$

for all $n > N$.

Suppose $a \neq b$, and let $c = |a - b| > 0$. Since 6.1 is valid for all positive rational numbers ϵ, we may choose $\epsilon = c/2$. Inequality 6.1 implies that $|a - b| < c$, contradicting the definition of c. Therefore $a = b$, so that the limit is unique. □

We note that the limit of the sequence (a) is a, since $|a - a| = 0 < \epsilon$ for any $\epsilon > 0$. This is a constant sequence, every term being equal to a.

Let (a_n) be a sequence on a set X and let (k_n) be any sequence on $\mathbb{N}$ such that $k_{n+1} > k_n$ for all $n \in \mathbb{N}$. The sequence (a_{k_n}) on X is called a **subsequence** of (a_n). For example, letting $(a_n) = (1/n)$ and $k_n = 2n - 1$ for each $n \in \mathbb{N}$, we obtain a subsequence $(1/(2n - 1))$ of $(1/n)$. Note that $k_1 \geq 1$ since (k_n) is a sequence on $\mathbb{N}$, and if $k_n \geq n$ for some $n \in \mathbb{N}$ then $k_{n+1} > k_n \geq n$, so that $k_{n+1} \geq n + 1$. It follows by induction that $k_n \geq n$ for all $n \in \mathbb{N}$.

Clearly every sequence has many subsequences. The next theorem shows that all the subsequences of a convergent sequence converge to the limit of the sequence.

Theorem 6.2 *Let (a_n) be a sequence on $\mathbb{Q}$ and suppose that*

$$\lim_{n \to \infty} a_n = a.$$

Then every subsequence (a_{k_n}) of (a_n) is convergent and

$$\lim_{n \to \infty} a_{k_n} = a.$$

Proof. Let (a_{k_n}) be a subsequence of (a_n). Choose $\epsilon > 0$. Since

$$\lim_{n \to \infty} a_n = a,$$

there exists an N such that $|a_n - a| < \epsilon$ for each $n > N$. Choose any $n > N$. Then $k_n > N$. Hence $|a_{k_n} - a| < \epsilon$ and so $\lim_{n \to \infty} a_{k_n} = a$. □

The contrapositive of Theorem 6.2 is particularly useful for establishing the divergence of a sequence. If there exists a divergent subsequence of (a_n), then (a_n) must also diverge; if there are two convergent subsequences of (a_n) with distinct limits, then (a_n) must diverge. For example, it is simple to show that the sequence $((-1)^n)$ is divergent using this result. Clearly the subsequences $((-1)^{2n})$ and $((-1)^{2n+1})$ converge to 1 and -1 respectively; therefore, the sequence $((-1)^n)$ must diverge because these limits are distinct. The next example is somewhat more involved.

Example 6.5 Consider the sequence (a_n) defined by

$$a_{n+1} = a_n - a_{n-1},$$

with $a_1 = 0$ and $a_2 = -1$. Here the behaviour of the terms as $n \to \infty$ is not immediately clear. The next ten terms of the sequence are $a_3 = -1, a_4 = 0, a_5 = a_6 = 1, a_7 = 0, a_8 = a_9 = -1, a_{10} = 0, a_{11} = a_{12} = 1$, and it seems that (a_n) diverges. To prove this conjecture the contrapositive of Theorem 6.2 can be applied to two suitable subsequences. Now

$$a_{6n+2} = a_{6n+1} - a_{6n} = (a_{6n} - a_{6n-1}) - a_{6n} = -a_{6n-1}$$
$$= -a_{6n-2} + a_{6n-3} = a_{6n-4} = a_{6(n-1)+2}$$
$$= -a_{6(n-1)-1} = \cdots,$$

and therefore the subsequences (a_{6n+2}) and (a_{6n-1}) converge to $a_8 = -1$ and $a_5 = 1$ respectively. Since $a_8 \neq a_5$, the sequence (a_n) must diverge.

△

A sequence (a_n) on $\mathbb{Q}$ is **bounded above** if there exists a number $\Lambda \in \mathbb{Q}$ such that $a_n \leq \Lambda$ for all $n \in \mathbb{N}$. Likewise (a_n) is **bounded below** if there is a number $\lambda \in \mathbb{Q}$ such that $a_n \geq \lambda$ for all $n \in \mathbb{N}$. The number Λ is called an **upper bound** for (a_n); the number λ is called a **lower bound** for (a_n).

A sequence is said to be **bounded** if it is bounded above and bounded below. Note that (a_n) is bounded if and only if there is some number Γ such that $|a_n| < \Gamma$ for all $n \in \mathbb{N}$. Indeed, if $|a_n| < \Gamma$ for all $n \in \mathbb{N}$ then $-\Gamma < a_n < \Gamma$ for all $n \in \mathbb{N}$, so that (a_n) is bounded above and below. For the converse we may assume that $\lambda < 0 < -\lambda < \Lambda$ and take any rational number $\Gamma > \Lambda$. If $\lambda \leq a_n \leq \Lambda$ then $-\Gamma < -\Lambda < \lambda \leq a_n \leq \Lambda < \Gamma$, so that $|a_n| < \Gamma$.

The sequence $(-n + 2)$, for example, is bounded above: any number Λ such that $\Lambda \geq 1$ provides an upper bound. The sequence $((-1)^n/n)$ is an example of a bounded sequence: any numbers Λ, λ such that $\Lambda \geq 1/2$ and $\lambda \leq -1$ provide upper and lower bounds respectively. The sequence $((-1)^n n)$ is neither bounded above nor bounded below.

It is clear that the sequence $((-1)^n)$ is bounded yet divergent, so that boundedness does not imply convergence. The converse, however, is true.

Theorem 6.3 *Any convergent sequence on $\mathbb{Q}$ is bounded.*

Proof. Let (a_n) be a convergent sequence on $\mathbb{Q}$. Suppose $a_n \to a$ as $n \to \infty$. Then for any $\epsilon > 0$ there exists an $N \in \mathbb{N}$ such that $|a_n - a| < \epsilon$

whenever $n > N$. Choose $\epsilon = 1$. Now,
$$|a_n| = |a_n - a + a| \le |a_n - a| + |a| < 1 + |a|$$
for all $n > N$. Let
$$\Gamma = \max\{|a_1|, |a_2|, \ldots, |a_N|, 1 + |a|\}.$$
Then $|a_n| \le \Gamma$ for all $n \in \mathbb{N}$ and thus (a_n) is bounded. $\square$

6.2 Cauchy Sequences

A feature of convergent sequences is not only that the terms approach a limit but that they become "close" to each other as $n \to \infty$ in a sense that will be made precise in this section. Consider, for example, the convergent sequence $(a_n) = (4 + (-1)^n/n^2)$. For any choice of $\epsilon > 0$ there exists an N such that $|a_n - 4| = 1/n^2 < \epsilon/2$ for each $n > N$. Suppose $m > n > N$. Then
$$|a_n - a_m| = |(4 + (-1)^n/n^2) - (4 + (-1)^m/m^2)|$$
$$\le \frac{1}{n^2} + \frac{1}{m^2} < \frac{2}{n^2} < \epsilon.$$
In other words, after some critical N, the difference between any two terms of (a_n) is less than ϵ.

A sequence (a_n) on $\mathbb{Q}$ is called a **Cauchy sequence** if for every positive rational number ϵ there exists some $N \in \mathbb{N}$ such that $|a_n - a_m| < \epsilon$ whenever $n > N$ and $m > N$. The condition requires all pairs of terms beyond a certain point to be close, not merely consecutive terms. Note that if (a_n) is a Cauchy sequence on $\mathbb{Q}$ then so is $(-a_n)$.

Theorem 6.4 *Every convergent sequence on $\mathbb{Q}$ is a Cauchy sequence.*

Proof. Let (a_n) be a convergent sequence on $\mathbb{Q}$ with $\lim_{n \to \infty} a_n = a$. Choose $\epsilon > 0$. It must be shown that there is an N such that $|a_n - a_m| < \epsilon$ whenever $n > N$ and $m > N$. Since $a_n \to a$ there exists an N such that $|a_n - a| < \epsilon/2$ for all $n > N$. Choose $n > N$ and $m > N$. Then
$$|a_n - a_m| = |a_n - a + a - a_m|$$
$$\le |a_n - a| + |a_m - a|$$
$$< \frac{\epsilon}{2} + \frac{\epsilon}{2}$$
$$= \epsilon,$$

and thus (a_n) is a Cauchy sequence. □

Though every convergent series is Cauchy, it is not yet clear if the converse is true. In fact, the converse is not true and this fact leads to the construction of the real number system. We will construct a divergent Cauchy sequence to prove this assertion in Section 6.4. Cauchy sequences are, however, bounded as the next theorem shows.

Theorem 6.5 *If a sequence on $\mathbb{Q}$ is Cauchy, then it is bounded.*

Proof. Let (a_n) be a Cauchy sequence on $\mathbb{Q}$. By definition there exists an $N \in \mathbb{N}$ such that $|a_m - a_n| < 1$ whenever $n > N$ and $m > N$. By taking $n = N + 1$ we obtain

$$a_{N+1} - 1 < a_m < a_{N+1} + 1$$

for all $m > N$, so that an upper bound is given by

$$\max\{a_1, a_2, \ldots, a_N, a_{N+1} + 1\}$$

and a lower bound is given by

$$\min\{a_1, a_2, \ldots, a_N, a_{N+1} - 1\}.$$

□

6.3 Operations with Cauchy Sequences

Given two sequences (a_n), (b_n) on $\mathbb{Q}$, addition and multiplication can be defined as follows:

(1) $(a_n) + (b_n) = (a_n + b_n)$;
(2) $(a_n)(b_n) = (a_n b_n)$.

It is natural to enquire whether addition and multiplication preserve the Cauchy property.

Theorem 6.6 *If (a_n) and (b_n) are Cauchy sequences on $\mathbb{Q}$ then so are $(a_n + b_n)$ and $(a_n b_n)$.*

Proof. It is first proved that $(a_n + b_n)$ is a Cauchy sequence. We must show that for any choice of $\epsilon > 0$ there is an N such that

$$|(a_n + b_n) - (a_m + b_m)| < \epsilon$$

for all $n, m > N$. By hypothesis (a_n) and (b_n) are Cauchy sequences; therefore, there exist numbers N_1 and N_2 such that

$$|a_n - a_m| < \frac{\epsilon}{2}$$

for all $n, m > N_1$ and

$$|b_n - b_m| < \frac{\epsilon}{2}$$

for all $n, m > N_2$. Let $N = \max\{N_1, N_2\}$. Thus, by the triangle inequality,

$$|(a_n + b_n) - (a_m + b_m)| \leq |a_n - a_m| + |b_n - b_m| < \frac{\epsilon}{2} + \frac{\epsilon}{2} = \epsilon$$

for all $n, m > N$ and consequently $(a_n + b_n)$ is Cauchy.

In order to prove that the sequence $(a_n b_n)$ is Cauchy it must be shown that for each $\epsilon > 0$ there is an N such that $|a_n b_n - a_m b_m| < \epsilon$ whenever $n, m > N$. Now

$$\begin{aligned} |a_n b_n - a_m b_m| &= |a_n b_n - a_m b_n + a_m b_n - a_m b_m| \\ &\leq |a_n b_n - a_m b_n| + |a_m b_n - a_m b_m| \\ &= |b_n||a_n - a_m| + |a_m||b_n - b_m|. \end{aligned}$$

The sequences (a_n) and (b_n) are Cauchy and therefore bounded. Let Λ_1 and Λ_2 be two positive rational numbers such that $|a_n| \leq \Lambda_1$ and $|b_n| \leq \Lambda_2$ for all $n \in \mathbb{N}$ and let $\Lambda = \max\{\Lambda_1, \Lambda_2\} > 0$. Since (a_n) is a Cauchy sequence there is an N_1 such that

$$|a_n - a_m| < \frac{\epsilon}{2\Lambda}$$

for all $n, m > N_1$. Similarly, there is an N_2 such that

$$|b_n - b_m| < \frac{\epsilon}{2\Lambda}$$

for all $n, m > N_2$. Let $N = \max\{N_1, N_2\}$. Therefore, for all $n, m > N$,

$$\begin{aligned} |a_n b_n - a_m b_m| &\leq |b_n||a_n - a_m| + |a_m||b_n - b_m| \\ &\leq \Lambda(|a_n - a_m| + |b_n - b_m|) \\ &< \Lambda\left(\frac{\epsilon}{2\Lambda} + \frac{\epsilon}{2\Lambda}\right) \\ &= \epsilon, \end{aligned}$$

and $(a_n b_n)$ is thus a Cauchy sequence. □

The set of all Cauchy sequences on $\mathbb{Q}$ will be denoted by $\mathbb{Q}^c$. Two sequences (a_n) and (b_n) in $\mathbb{Q}^c$ are said to be **equivalent** if for every positive rational number ϵ there exists a natural number N such that $|a_n - b_n| < \epsilon$ for all $n > N$. We write $(a_n) \sim (b_n)$ if this condition holds. Loosely speaking, two Cauchy sequences are equivalent if beyond a finite number of terms they are termwise close. Note that $(a_n) \sim (b_n)$ if and only if $(a_n - b_n) \sim (0)$.

Theorem 6.7 *The relation $\sim$ is an equivalence relation on $\mathbb{Q}^c$.*

Proof. It is immediate from the definition of equivalence that $(a_n) \sim (a_n)$ for all $(a_n) \in \mathbb{Q}^c$. Moreover, since $|a_n - b_n| = |b_n - a_n|$ it is clear that $(a_n) \sim (b_n)$ implies that $(b_n) \sim (a_n)$ for all $(a_n), (b_n) \in \mathbb{Q}^c$. It remains thus to show that the relation is transitive. Suppose $(a_n) \sim (b_n)$ and $(b_n) \sim (c_n)$ for $(a_n), (b_n), (c_n) \in \mathbb{Q}^c$. It is required to show that for any $\epsilon > 0$ there is an N such that $|a_n - c_n| < \epsilon$ for all $n > N$. Choose $\epsilon > 0$. The triangle inequality yields

$$|a_n - c_n| \leq |a_n - b_n| + |b_n - c_n|.$$

Since $(a_n) \sim (b_n)$, there is an N_1 such that $|a_n - b_n| < \epsilon/2$ for all $n > N_1$. Similarly, since $(b_n) \sim (c_n)$, there is an N_2 such that $|b_n - c_n| < \epsilon/2$ for all $n > N_2$. Taking $N = \max\{N_1, N_2\}$ we see that $|a_n - c_n| < \epsilon/2 + \epsilon/2 = \epsilon$ for all $n > N$ and thus $(a_n) \sim (c_n)$. □

Example 6.6 Prove that

$$\left(\frac{1}{n}\right) \sim (0).$$

Solution: Choose $\epsilon > 0$ and let $N \geq 1/\epsilon$. Now $|1/n - 0| = 1/n$, so that if $n > N \geq 1/\epsilon$ then $|1/n - 0| < \epsilon$. Hence the equivalence follows.

△

Example 6.7 Prove that

$$\left(4 - \frac{1}{n}\right) \sim \left(4 + \frac{1}{n^2}\right).$$

Solution: Choose $\epsilon > 0$. Let $a_n = 4 - 1/n$ and $b_n = 4 + 1/n^2$ for all $n \in \mathbb{N}$. Since

$$|a_n - b_n| = \frac{1}{n}\left(1 + \frac{1}{n}\right) \leq \frac{2}{n},$$

we have $|a_n - b_n| < \epsilon$ for all $n > 2/\epsilon$. Thus the equivalence follows. △

The equivalence relation allows one to partition the set $\mathbb{Q}^c$ into equivalence classes. Every sequence in $\mathbb{Q}^c$ belongs to some class but no sequence belongs to two distinct classes. If (a_n) belongs to a particular class, then all sequences (b_n) equivalent to (a_n) belong to that class and any member of the class can be used to represent the class. The class containing the sequence (a_n) will be denoted by $[(a_n)]$. Since $(a_n) \sim (b_n)$ if and only if $(a_n - b_n) \sim (0)$, it follows that $[(a_n)] = [(b_n)]$ if and only if $[(a_n - b_n)] = [(0)]$.

Theorem 6.4 shows that if (a_n) is a sequence on $\mathbb{Q}$ such that $a_n \to a$ as $n \to \infty$ then $(a_n) \in \mathbb{Q}^c$. Clearly, the sequence (a) whose terms are all equal to $a \in \mathbb{Q}$ is in $\mathbb{Q}^c$ and the definitions of convergence and equivalence show that (a_n) converges to a if and only if $(a_n) \sim (a)$. The sequences (a_n) and (a) are thus in the same equivalence class. There are no other constant sequences in this class: if a and b are distinct rational numbers then $|a - b| > |a - b|/2$, and so (a) and (b) cannot be equivalent. Note that for any $a \in \mathbb{Q}$, the equivalence class $[(a)]$ does not contain divergent sequences. This observation is evident from the definition of equivalence: any sequence $(b_n) \in [(a)]$ must satisfy $\lim_{n\to\infty} b_n = a$. For a constant sequence (a) we usually write $[a]$ instead of $[(a)]$.

If $[(a_n)]$ and $[(b_n)]$ are distinct classes, questions arise concerning the equivalence classes formed under addition or multiplication of representatives from these classes. For example, suppose $(a_n) \sim (c_n)$ and $(b_n) \sim (d_n)$. Is $(a_n + b_n) \sim (c_n + d_n)$? The next theorem addresses this concern.

Theorem 6.8 *Let $(a_n), (b_n), (c_n), (d_n) \in \mathbb{Q}^c$ and suppose that $(a_n) \sim (c_n)$ and $(b_n) \sim (d_n)$. Then $(a_n + b_n) \sim (c_n + d_n)$ and $(a_n b_n) \sim (c_n d_n)$.*

Proof. It will first be shown that $(a_n + b_n) \sim (c_n + d_n)$. Choose any $\epsilon > 0$. By definition, there exist natural numbers N_1, N_2 such that

$$|a_n - c_n| < \frac{\epsilon}{2}$$

for all $n > N_1$, and
$$|b_n - d_n| < \frac{\epsilon}{2}$$
for all $n > N_2$. Let $N = \max\{N_1, N_2\}$. Then
$$\begin{aligned}|(a_n + b_n) - (c_n + d_n)| &\leq |a_n - c_n| + |b_n - d_n| \\ &< \frac{\epsilon}{2} + \frac{\epsilon}{2} \\ &= \epsilon,\end{aligned}$$
as required.

To show that $(a_n b_n) \sim (c_n d_n)$, observe that
$$\begin{aligned}|a_n b_n - c_n d_n| &= |a_n b_n - a_n d_n + a_n d_n - c_n d_n| \\ &\leq |a_n||b_n - d_n| + |d_n||a_n - c_n|,\end{aligned}$$
and since $(a_n), (d_n) \in \mathbb{Q}^c$, there exist positive constants Γ_1 and Γ_2 such that $|a_n| \leq \Gamma_1$ and $|d_n| \leq \Gamma_2$ for all $n \in \mathbb{N}$. Moreover, for any $\epsilon_1 > 0$ and $\epsilon_2 > 0$ there exist natural numbers N_1, N_2 such that $|a_n - c_n| < \epsilon_1$ for all $n > N_1$ and $|b_n - d_n| < \epsilon_2$ for all $n > N_2$. Therefore,
$$|a_n b_n - c_n d_n| < \Gamma_1 \epsilon_1 + \Gamma_2 \epsilon_2,$$
provided $n > \max\{N_1, N_2\}$. The inequality $|a_n b_n - c_n d_n| < \epsilon$ thus follows upon choosing $\epsilon_1 = \epsilon/2\Gamma_1$ and $\epsilon_2 = \epsilon/2\Gamma_2$. □

6.4 A Divergent Cauchy Sequence

In this section we show that there exists a divergent Cauchy sequence. This observation is crucial because it highlights the inadequacy of the rational number system and provides a key for an extension. Our strategy for finding such a sequence is first to show that there is no solution x, among the rational numbers, to the equation $x^2 = 2$ and then to establish the existence of a Cauchy sequence (a_n) such that $a_n^2 \to 2$ as $n \to \infty$.

Theorem 6.9 *There does not exist $x \in \mathbb{Q}$ such that $x^2 = 2$.*

Proof. Suppose there is an $x \in \mathbb{Q}$ such that $x^2 = 2$. Since x is rational we may write $x = a/b$ where $a, b \in \mathbb{Z}$ and $b > 0$. Suppose a and b are chosen to minimize b. Now $x^2 = a^2/b^2 = 2$, and therefore $2b^2 = a^2$. This implies that a^2 is even. Hence a must also be even, for if a were odd, then we should have $a = 2f + 1$ for some integer f, so that $a^2 = 4f^2 + 4f + 1$, which is also odd. Let $a = 2c$, where $c \in \mathbb{Z}$. Then $4c^2 = a^2 = 2b^2$ so that $b^2 = 2c^2$ and so b is even. Thus there is a positive integer d such that $b = 2d$. Clearly $d < b$, and $x = a/b = c/d$. This result contradicts the choice of a and b and thus establishes the theorem. □

Lemma 6.10 *There exists a sequence (a_n), of positive rational numbers, such that $a_n^2 \to 2$ as $n \to \infty$.*

Proof. Choose $\epsilon > 0$. Theorem 5.18 implies that for any $n \in \mathbb{N}$ there exists a positive number a_n such that $2 < a_n^2 < 2 + 1/n$. Let (a_n) be a sequence on $\mathbb{Q}$ where each a_n has this property and let $N \geq 1/\epsilon$. Then $2 < a_n^2 < 2 + \epsilon$ for all $n > N$ and consequently $|a_n^2 - 2| < \epsilon$ for all $n > N$. Therefore, $a_n^2 \to 2$ as $n \to \infty$. □

Lemma 6.11 *The sequence (a_n) of Lemma 6.10 is Cauchy.*

Proof. Suppose that (a_n) is not a Cauchy sequence. Then there is an $\epsilon > 0$ such that for all $N \in \mathbb{N}$ there exists a pair n, m of natural numbers, greater than N, for which $|a_n - a_m| \geq \epsilon$. Let n and m be such a pair. Then

$$|a_n^2 - a_m^2| = |(a_n - a_m)(a_n + a_m)|$$
$$= |a_n - a_m||a_n + a_m|$$
$$\geq \epsilon(a_n + a_m)$$
$$> 2\epsilon,$$

since $a_n > 1$ and $a_m > 1$.

Lemma 6.10 shows that the sequence (a_n^2) is convergent and hence by Theorem 6.4 it must also be Cauchy. Therefore, there is an $N_1 \in \mathbb{N}$ such that for any $k > N_1$ and $j > N_1$ the inequality $|a_k^2 - a_j^2| < 2\epsilon$ is satisfied. But for any $N \in \mathbb{N}$ we have established the existence of a pair n, m such that $n > N$, $m > N$ and $|a_n^2 - a_m^2| > 2\epsilon$. This contradiction establishes that (a_n) is a Cauchy sequence. □

Lemma 6.12 *Let (a_n) be a sequence on $\mathbb{Q}$. If $a_n \to a$ as $n \to \infty$ then $a_n^2 \to a^2$ as $n \to \infty$.*

Proof. Choose $\epsilon > 0$. Now,

$$|a_n^2 - a^2| = |a_n - a||a_n + a| \leq |a_n - a|(|a_n| + |a|),$$

and since (a_n) is convergent Theorems 6.4 and 6.5 imply that it is bounded. Therefore there is some positive rational number Γ such that

$$|a_n^2 - a^2| \leq |a_n - a|(\Gamma + |a|).$$

Moreover, because $a_n \to a$ there is an $N \in \mathbb{N}$ such that

$$|a_n - a| < \frac{\epsilon}{\Gamma + |a|}$$

for all $n > N$; consequently, $|a_n^2 - a^2| < \epsilon$ whenever $n > N$ and thus the sequence (a_n^2) converges to a^2. □

Theorem 6.13 *The sequence (a_n) of Lemma 6.10 is divergent in $\mathbb{Q}$.*

Proof. Suppose $a_n \to a$ as $n \to \infty$, where $a \in \mathbb{Q}$. Lemma 6.12 implies that $a_n^2 \to a^2$ as $n \to \infty$. But $a_n^2 \to 2$ and therefore $a^2 = 2$, contradicting Theorem 6.9. □

Exercises 6

(1) Use the definition of convergence to prove that the sequence (a_n) converges, where a_n is given by

   (a) $\dfrac{n+2}{n^2}$,

   (b) $\dfrac{1+(-1)^n}{n}$,

   (c) $\dfrac{n(n+1)(n+2)}{(n+3)(n+4)(n+5)}$.

(2) Determine whether the following sequences converge:

   (a) $\left(\dfrac{1}{n+(-1)^n}\right)$,

   (b) $\left(\dfrac{1+(-1)^n}{n}\right)$,

   (c) $\left(\dfrac{1}{1+n^2(1+(-1)^n)}\right)$,

   (d) $\left(n - \dfrac{n^2+1}{n+1}\right)$.

(3) Suppose that $a_n \to a$ as $n \to \infty$ and $b_n \to b$ as $n \to \infty$.

   (a) Prove that $a_n + b_n \to a + b$ as $n \to \infty$.
   (b) Prove that $a_n b_n \to ab$ as $n \to \infty$.

(4) Let (a_n) be a bounded sequence and (b_n) a Cauchy sequence. Prove that $(a_n b_n)$ is a bounded sequence.

(5) Suppose that (a_n) and (a_n+b_n) are Cauchy sequences. Prove that (b_n) is also a Cauchy sequence.

(6) Given that $(a_n) \sim (b_n)$, prove that $(a_n^2) \sim (b_n^2)$.

Chapter 7

The Real Number System

The integer and rational number systems were constructed via equivalence classes and it should occasion little surprise that the real number system is constructed in a similar manner. The mathematical apparatus required to define the requisite equivalence relation is perhaps more complicated than that for earlier systems, but the underlying strategy is essentially the same.

7.1 The Real Numbers

Let $[\mathbb{Q}^c]$ denote the set of equivalence classes on $\mathbb{Q}^c$ under the equivalence relation of the previous chapter. A notable feature of the set $[\mathbb{Q}^c]$ from the preceding discussion is that some equivalence classes are represented by constant sequences. It follows from Theorem 6.4 that every convergent sequence on $\mathbb{Q}$ belongs to a class having a constant sequence as a representative. On the other hand, the results of Section 6.4 show that there exists at least one equivalence class which does not have such a representative. The set $[\mathbb{Q}^c]$ can thus be partitioned into two nonempty sets:

(1) $\mathbb{R}_\mathbb{Q} = \{[a] : a \in \mathbb{Q}\}$;
(2) $\mathbb{R}_I = [\mathbb{Q}^c] - \mathbb{R}_\mathbb{Q}$.

The members of $\mathbb{R}_I$ are called **irrational numbers**. Let $\mathbb{R} = \mathbb{R}_\mathbb{Q} \cup \mathbb{R}_I = [\mathbb{Q}^c]$ and denote $[0]$ and $[1]$ by 0 and 1 respectively. The induced operations of addition and multiplication will be denoted by $+$ and $\cdot$ respectively. In other words we define

$$[(a_n)] + [(b_n)] = [(a_n + b_n)]$$

and
$$[(a_n)][(b_n)] = [(a_n b_n)].$$

It follows from Theorem 6.6 and Theorem 6.8 that these are well defined operations on $\mathbb{R}$.

Let us define $\Phi(a) = [a]$ for each $a \in \mathbb{Q}$. Then Φ is a bijection from $\mathbb{Q}$ onto $\mathbb{R}_\mathbb{Q}$. Moreover if $(a,b) \in \mathbb{Q} \times \mathbb{Q}$ then $\Phi(a) + \Phi(b) = [a] + [b] = [a+b] = \Phi(a+b)$. Similarly $\Phi(a)\Phi(b) = \Phi(ab)$. Thus $\mathbb{R}_\mathbb{Q}$ may be identified with $\mathbb{Q}$. In particular, note that $\Phi(0) = [0]$ and similarly $\Phi(1) = [1]$. Thus 0 and 1 in $\mathbb{R}$ are reconciled with the corresponding rational numbers. We are now free to write a instead of $[a]$ when $a \in \mathbb{Q}$, but for the sake of clarity we continue to use the latter notation on occasion.

Theorem 7.1 $(\mathbb{R}, +, \cdot, 0, 1)$ *is a commutative ring with unity.*

Proof. That $(\mathbb{R}, +, \cdot, 0, 1)$ is a commutative ring with unity follows immediately from Theorem 5.4 (which asserts the same result for $(\mathbb{Q}, +, \cdot, 0, 1)$), the definitions of $+$ and $\cdot$, and the fact that 0 and 1 are distinct rational numbers. For instance, we have

$$[(a_n)] + [(b_n)] = [(a_n + b_n)] = [(b_n + a_n)] = [(b_n)] + [(a_n)].$$

Note that $-[(a_n)] = [(-a_n)]$ since

$$[(a_n)] + [(-a_n)] = [(a_n - a_n)] = [0] = 0.$$

The proofs of the other axioms are similar. $\square$

Note also that

$$\begin{aligned}
[(a_n)] - [(b_n)] &= [(a_n)] + (-[(b_n)]) \\
&= [(a_n)] + [(-b_n)] \\
&= [(a_n - b_n)].
\end{aligned}$$

It is worth noting that the equation $x^2 = 2$ can be solved in the real number system.

Corollary 7.2 *There exists an $x \in \mathbb{R}$ such that $x^2 = 2$.*

Proof. By Lemmas 6.10 and 6.11 there is a Cauchy sequence (a_n) on $\mathbb{Q}$ such that $a_n^2 \to 2$ as $n \to \infty$. Let $x = [(a_n)]$. Then

$$x^2 = [(a_n)]^2 = [(a_n^2)] = [2] = 2.$$

$\square$

Evidently, any solution to the equation $x^2 = 2$ is an irrational number.

Lemma 7.3 *Suppose $[(a_n)] \neq 0$. Then there exist a positive rational number λ and a natural number N such that $|a_n| > \lambda$ for all $n > N$.*

Proof. Suppose the lemma fails for some Cauchy sequence (a_n) such that $[(a_n)] \neq 0$. Then for every rational $\lambda > 0$ and every $N \in \mathbb{N}$ there exists $n > N$ such that $|a_n| \leq \lambda$.

We shall derive a contradiction by proving that $a_n \to 0$. Choose $\epsilon > 0$. Since $(a_n) \in \mathbb{Q}^c$ there is an $N \in \mathbb{N}$ such that $|a_n - a_m| < \epsilon/2$ whenever $m, n > N$. The triangle inequality yields

$$|a_n| = |a_m + a_n - a_m|$$
$$\leq |a_m| + |a_n - a_m|$$
$$< |a_m| + \epsilon/2$$

whenever $m, n > N$. Choose $m > N$ such that $|a_m| \leq \epsilon/2$ (taking $\lambda = \epsilon/2$ in the previous paragraph). Then $|a_n| < \epsilon/2 + \epsilon/2 = \epsilon$ for all $n > N$. Consequently, $a_n \to 0$ as $n \to \infty$ in contradiction to the assumption that $[(a_n)] \neq 0$. $\square$

A **null sequence** is any sequence in $[0]$. The above lemma shows that if (a_n) is not a null sequence then it can be "bounded away" from zero. In other words, there is a *positive* lower bound on $|a_n|$ for n sufficiently large. This result can be used to prove that a multiplicative inverse exists and thus division can be defined on $\mathbb{R} - \{0\}$.

Theorem 7.4 *If $s = [(a_n)] \in \mathbb{R} - \{0\}$, then there exists an $r \in \mathbb{R}$ such that $rs = 1$.*

Proof. Suppose $s = [(a_n)] \in \mathbb{R} - \{0\}$. Then, by Lemma 7.3 there exist a positive rational number λ and a natural number N_1 such that $|a_n| > \lambda$ for each $n > N_1$. Therefore, by Theorem 5.5, for all $n > N_1$ each a_n has a multiplicative inverse $1/a_n \in \mathbb{Q}$. Note that

$$\frac{1}{a_n} \leq \frac{1}{|a_n|} < \frac{1}{\lambda}.$$

Define a sequence (b_n) by letting $b_n = 1$ for each $n \leq N_1$ and $b_n = 1/a_n$ for each $n > N_1$. For all m and n greater than N_1 we have

$$|b_n - b_m| = \left| \frac{1}{a_n} - \frac{1}{a_m} \right|$$

$$= \frac{1}{|a_n||a_m|} |a_m - a_n|$$

$$\leq \frac{|a_m - a_n|}{\lambda^2}.$$

Since $(a_n) \in \mathbb{Q}^c$, for each $\epsilon > 0$ there exists N_2 such that

$$|a_m - a_n| < \lambda^2 \epsilon$$

whenever $m, n > N_2$. Let $N = \max\{N_1, N_2\}$. Then $|b_n - b_m| < \epsilon$ whenever $m > N$ and $n > N$, so that (b_n) is a Cauchy sequence. Let $r = [(b_n)]$; then,

$$rs = [(a_n)][(b_n)] = [(a_n b_n)] = [1] = 1.$$

$\square$

Theorems 7.1 and 7.4 imply that $(\mathbb{R}, +, \cdot, 0, 1)$ is a field. Therefore the general results of Chapter 5 can be used. For example, Theorem 5.6 shows that $(\mathbb{R}, +, \cdot, 0, 1)$ is an integral domain. The theorems concerning exponentiation detailed in Section 5.4 are all valid for $(\mathbb{R}, +, \cdot, 0, 1)$. Here, $r^n = \prod_{k=1}^{n} r$ for any $r \in \mathbb{R}$ and $n \in \mathbb{N}$, in agreement with the earlier definition of exponentiation. We also define $r^0 = 1$ if $r \in \mathbb{R} - \{0\}$.

Following the pattern of the development of the previous number systems, the next major step is to generalize the concept of order. This concept has already been extended to the rational numbers and, in the spirit of the earlier generalizations, any such concept must express a direct extension of the existing version: if the rational numbers a and b satisfy $a < b$ the new refinement must yield $[a] < [b]$. Central to this generalization is an extension of the concept of a positive number. A sequence $(a_n) \in \mathbb{Q}^c$ is **positive** if there exists some positive rational number ϵ and some $N \in \mathbb{N}$ such that $a_n > \epsilon$ for all $n > N$. Note that a constant sequence (a) is positive if and only if $a > 0$. The next lemma shows that the property of being positive permeates the entire equivalence class associated with a positive sequence.

Lemma 7.5 *If $(a_n) \in \mathbb{Q}^c$ is positive then so is every $(b_n) \in [(a_n)]$.*

Proof. Suppose (a_n) is positive and $(b_n) \sim (a_n)$. Since (a_n) is positive there exist a positive rational number ϵ and a natural number N_1 such that $a_n > \epsilon$ for all $n > N_1$; since $(b_n) \sim (a_n)$, there exists a natural number N_2

such that $|a_n - b_n| < \epsilon/2$ whenever $n > N_2$. Let $N = \max\{N_1, N_2\}$. For all $n > N$ we have $a_n - b_n < \epsilon/2$, so that $b_n > a_n - \epsilon/2 > \epsilon - \epsilon/2 = \epsilon/2 > 0$. Thus (b_n) is positive. □

Some easy consequences of the definitions of a positive Cauchy sequence and positive rational numbers are contained in the following lemma.

Lemma 7.6

(a) *Let $(a_n) \in \mathbb{Q}^c$ and $(b_n) \in \mathbb{Q}^c$ be positive sequences. Then the sequences $(a_n + b_n)$ and $(a_n b_n)$ are positive.*

(b) *For any sequence $(a_n) \in \mathbb{Q}^c$ precisely one of the following is true:*

(i) (a_n) *is positive;*
(ii) (a_n) *is a null sequence;*
(iii) $(-a_n)$ *is positive.*

Proof. (a) Since (a_n) is positive there exist a positive rational number ϵ_1 and an $N_1 \in \mathbb{N}$ such that $a_n > \epsilon_1$ for all $n > N_1$. Similarly there exist a positive rational number ϵ_2 and an $N_2 \in \mathbb{N}$ such that $b_n > \epsilon_2$ for all $n > N_2$. Let $N = \max\{N_1, N_2\}$, and choose $n > N$. Since $n > N_1$ we have $a_n > \epsilon_1$. Similarly $b_n > \epsilon_2$, and so $a_n + b_n > \epsilon_1 + \epsilon_2 > 0$ and $a_n b_n > \epsilon_1 \epsilon_2 > 0$. Hence $(a_n + b_n)$ and $(a_n b_n)$ are positive.

(b) Suppose (a_n) is non-null. By Lemma 7.3 there exist a positive rational number ϵ and an $N_1 \in \mathbb{N}$ such that $|a_n| > \epsilon$ for all $n > N_1$. Hence either $a_n > \epsilon$ or $-a_n > \epsilon$ for all $n > N_1$. By the Cauchy property there exists $N_2 \in \mathbb{N}$ such that $|a_n - a_m| < \epsilon$ whenever $n > N_2$ and $m > N_2$. Let $N = \max\{N_1, N_2\}$. If there exist $m, n \geq N$ such that $a_n > \epsilon$ and $-a_m > \epsilon$, then $a_n - a_m > 2\epsilon$, in violation of the Cauchy property. It follows that either $a_n > \epsilon$ for all $n > N$ or $-a_n > \epsilon$ for all $n > N$. In other words, either (a_n) or $(-a_n)$ is positive. We conclude that at least one of (i), (ii) or (iii) holds.

Next, suppose that (a_n) is null. For every positive rational number ϵ there exists $N \in \mathbb{N}$ such that $|a_n| < \epsilon$ for all $n > N$. Hence (a_n) cannot be positive. Neither can $(-a_n)$ be positive, for if $-a_n > \epsilon > 0$ then $|a_n| > \epsilon$.

It remains only to show that (a_n) and $(-a_n)$ cannot both be positive. If (a_n) is positive then for some positive rational number ϵ_1 and $N_1 \in \mathbb{N}$ we have $a_n > \epsilon_1$ for all $n > N_1$. If $(-a_n)$ is also positive, then for some positive rational number ϵ_2 and $N_2 \in \mathbb{N}$ we have $-a_n > \epsilon_2$ for all $n > N_2$. Let $\epsilon = \min\{\epsilon_1, \epsilon_2\}$ and $N = \max\{N_1, N_2\}$, and choose $n > N$. Then

$n > N_1$ and so $a_n > \epsilon_1 \geq \epsilon > 0 > -a_n$. Similarly $n > N_2$, and we reach the contradiction that $-a_n > \epsilon_2 \geq \epsilon$. □

Lemma 7.5 indicates that the property of a sequence being positive is really a property of the equivalence class to which it belongs, and this circumstance allows one to define a "positive" equivalence class. Specifically, $[(a_n)]$ is **positive** if (a_n) is positive; $[(a_n)]$ is **negative** if $(-a_n)$ is positive. The relation $<$ is defined on $\mathbb{R}$ as follows: we write $[(a_n)] < [(b_n)]$ or $[(b_n)] > [(a_n)]$ if $[(b_n - a_n)]$ is positive. Note that Theorem 6.6 and Lemma 7.5 ensure that this relation is well defined on $\mathbb{R}$. Moreover, it is an extension of the concept of order developed for the rational number system. Indeed, if a and b are rational numbers such that $a < b$, then $b - a > 0$, so that $(b - a)$ is a positive sequence and $[a] < [b]$. This argument is reversible, so that in fact $[a] < [b]$ if and only if $a < b$. Note also that $[(a_n)] > 0$ if and only if $[(a_n)]$ is positive, and therefore that $[(a_n)] < [(b_n)]$ if and only if $[(b_n)] - [(a_n)] = [(b_n - a_n)] > 0$. Similarly $[(a_n)] < 0$ if and only if $[(a_n)]$ is negative, and $[(a_n)] > [(b_n)]$ if and only if $[(a_n)] - [(b_n)] > 0$. Moreover $[(a_n)] \not< [(a_n)]$ since $[(a_n - a_n)] = [0] = 0$, which is not positive.

If $r, s \in \mathbb{R}$ and $r < s$ then r is said to be **less** than s, and s is said to be **greater** than r, in agreement with the earlier usages of these terms. We also write $r \leq s$ or $s \geq r$ if either $r < s$ or $r = s$. Thus $[(a_n)] \geq 0$ if there exists $N \in \mathbb{N}$ such that $a_n \geq 0$ for all $n > N$, for if $[(a_n)] < 0$ then $[(-a_n)] > 0$ and we have the contradiction that there is a rational number $\epsilon > 0$ such that $-a_n > \epsilon$ (so that $a_n < -\epsilon < 0$) for all $n \in \mathbb{N}$ greater than some natural number N_1.

It can now be verified that $(\mathbb{R}, +, \cdot, 0, 1, <)$ is an ordered integral domain (cf. Section 4.6). The axioms for an ordered integral domain can be checked exactly as in the case of the integers and rational numbers. Since $(\mathbb{R}, +, \cdot, 0, 1)$ is a field, $(\mathbb{R}, +, \cdot, 0, 1, <)$ is an ordered field. This ordered field will be called the **real number system**.

7.2 Sequences on $\mathbb{R}$

The real number system is an ordered integral domain and therefore the absolute value of any real number is already defined (cf. Section 4.7). Note that if $q \in \mathbb{Q}$ and $r = [q] \in \mathbb{R}$, then $r^2 = [q^2]$. Thus

$$|r| = [|q|],$$

since $[|q|] \geq 0$ and $[|q|]^2 = [|q|^2] = [q^2] = r^2$, and so the absolute value for an element of $\mathbb{R}$ is an extension of the absolute value for an element of $\mathbb{Q}$. If $(a_n) \in \mathbb{Q}^c$ and $r = [(a_n)]$, then

$$|r|^2 = r^2 = [(a_n)]^2 = [(a_n^2)] = [(|a_n|^2)] = [(|a_n|)]^2,$$

and so $|r| = [(|a_n|)]$ since $[(|a_n|)] \geq 0$. Other properties of the absolute value which rely only on $(\mathbb{R}, +, \cdot, 0, 1, <)$ being an ordered integral domain can be deduced immediately from the general results of Section 4.7. In particular, if r and s are real numbers then:

(a) $|r| \geq 0$;
(b) $|rs| = |r||s|$;
(c) $||r| - |s|| \leq |r + s| \leq |r| + |s|$.

The definition of convergence for sequences on $\mathbb{R}$ is formally the same as that for sequences on $\mathbb{Q}$ but ϵ may now be any positive real number. A sequence (r_n) on $\mathbb{R}$ is said to **converge** to a number $r \in \mathbb{R}$ if and only if for every positive real number ϵ there exists a natural number N such that $|r_n - r| < \epsilon$ whenever $n > N$.

It is intuitively clear that a sequence on $\mathbb{Q}$ can also be regarded as a sequence on $\mathbb{R}$ because each term of the sequence on $\mathbb{Q}$ is a rational number and can be identified with a real number via its equivalence class. Let (a_n) be a sequence on $\mathbb{Q}$ and for each $n \in \mathbb{N}$ let $r_n = [a_n]$. (Each r_n is therefore the equivalence class of a constant sequence.) The sequence (r_n) on $\mathbb{R}$ is said to be **induced** by the sequence (a_n). Note that this sequence is on $\mathbb{R}_\mathbb{Q}$. Conversely, any sequence on $\mathbb{R}_\mathbb{Q}$ is induced by a sequence on $\mathbb{Q}$. An immediate question concerns whether convergent sequences on $\mathbb{Q}$ induce convergent sequences on $\mathbb{R}$. The following lemma is of use in resolving this matter.

Lemma 7.7 *For any positive real number ϵ there exists $\eta \in \mathbb{Q}$ such that $0 < \eta < \epsilon$.*

Proof. Let $\epsilon = [(\epsilon_n)]$, where $(\epsilon_n) \in \mathbb{Q}^c$. Since (ϵ_n) is positive there exists a positive rational number λ and a natural number N such that $\epsilon_n > \lambda$ for all $n > N$. The density property of the rationals (Theorem 5.13) implies that there is an $\eta \in \mathbb{Q}$ such that $0 < \eta < \lambda$. Therefore $\epsilon_n - \eta > \epsilon_n - \lambda > 0$ for all $n > N$. Thus $(\epsilon_n - \eta)$ is a positive sequence, so that $[(\epsilon_n)] > [\eta]$. The lemma follows. $\square$

One application of this lemma is to show that the definition of convergence for sequences on $\mathbb{R}$ is an extension of that for sequences on $\mathbb{Q}$.

Lemma 7.8 *Let (q_n) be a sequence on $\mathbb{Q}$ and suppose that $q_n \to q$ in $\mathbb{Q}$ as $n \to \infty$. Then the induced sequence (r_n) on $\mathbb{R}$ converges to $[q]$.*

Proof. Let $r_n = [q_n]$ for each $n \in \mathbb{N}$ and let $r = [q]$. Choose any positive real number ϵ. By Lemma 7.7 there exists a positive rational number η such that $[\eta] < \epsilon$. Since $q_n \to q$, there is an $N \in \mathbb{N}$ such that $|q_n - q| < \eta$ for all $n > N$. Thus for each such n we have

$$|r_n - r| = |[q_n] - [q]| = [|q_n - q|] < [\eta] < \epsilon.$$

The sequence (r_n) therefore converges to r. □

The definition of a Cauchy sequence on $\mathbb{R}$ is formally the same as that for a Cauchy sequence on $\mathbb{Q}$. Specifically, a sequence (r_n) on $\mathbb{R}$ is said to be a **Cauchy sequence** if and only if for any positive real number ϵ there is an $N \in \mathbb{N}$ such that $|r_n - r_m| < \epsilon$ for all $n > N$ and $m > N$.

Lemma 7.9 *Any Cauchy sequence on $\mathbb{R}_\mathbb{Q}$ is induced by a Cauchy sequence on $\mathbb{Q}$.*

Proof. A Cauchy sequence (r_n) on $\mathbb{R}_\mathbb{Q}$ must be induced by some sequence (a_n) on $\mathbb{Q}$. In other words, $r_n = [a_n]$ for each $n \in \mathbb{N}$. We must prove that the sequence (a_n) is Cauchy. Choose a positive rational number ϵ. As (r_n) is Cauchy, there exists $N \in \mathbb{N}$ such that $|r_m - r_n| < [\epsilon]$ for all $m > N$ and $n > N$. Choose $m > N$ and $n > N$. Since

$$|r_m - r_n| = |[a_m] - [a_n]| = |[a_m - a_n]| = [|a_m - a_n|],$$

we have $[|a_m - a_n|] < [\epsilon]$, and so $|a_m - a_n| < \epsilon$, as required. □

Lemma 7.10 *If (a_n) is a Cauchy sequence on $\mathbb{Q}$, then the induced sequence (r_n) is a Cauchy sequence on $\mathbb{R}$.*

Proof. Choose any positive real number ϵ. By Lemma 7.7 there is a positive rational number η such that $[\eta] = \eta < \epsilon$. Since (a_n) is a Cauchy sequence there exists an $N \in \mathbb{N}$ such that $|a_n - a_m| < \eta$ for all $n > N$ and $m > N$. Now the terms in the induced sequence are given by $r_n = [a_n]$, and for each $m > N$ and $n > N$ we have

$$|r_n - r_m| = [|a_n - a_m|],$$

so that

$$\epsilon - |r_n - r_m| > [\eta] - [|a_n - a_m|] = [\eta - |a_n - a_m|].$$

For any $n, m > N$, $\eta - |a_n - a_m| > 0$. Hence $[\eta - |a_n - a_m|] > 0$, and so $|r_n - r_m| < \epsilon$. □

Theorem 6.13 shows that not all Cauchy sequences on $\mathbb{Q}$ converge. Certainly Lemma 7.8 shows that convergent Cauchy sequences on $\mathbb{Q}$ induce convergent sequences on $\mathbb{R}$, but it may be that a divergent Cauchy sequence on $\mathbb{Q}$ such as that of Theorem 6.13 induces a *convergent* sequence on $\mathbb{R}$. Indeed, this question lies at the very core of the original motivation for constructing the real number system. The following theorem shows that *any* Cauchy sequence on $\mathbb{Q}$ induces a convergent sequence on $\mathbb{R}$.

Theorem 7.11 *Let $(a_n) \in \mathbb{Q}^c$ and let $r = [(a_n)]$. Let (r_n) be the sequence on $\mathbb{R}_\mathbb{Q}$ induced by (a_n). Then (r_n) is convergent and $r_n \to r$ as $n \to \infty$.*

Proof. Choose a real number $\epsilon > 0$. It is required to show that there exists $N \in \mathbb{N}$ such that

$$|r_m - r| < \epsilon$$

for all $m > N$. Fix $m > N$. Now,

$$|r_m - r| = |[a_m] - [(a_n)]|$$
$$= |[(a_m - a_n)]|$$
$$= [(|a_n - a_m|)],$$

and Lemma 7.7 implies that there exists some positive rational number η such that $[\eta] < \epsilon$. Therefore

$$\epsilon - |r_m - r| > [\eta] - [(|a_n - a_m|)]$$
$$= [(\eta - |a_n - a_m|)].$$

Now $(a_n) \in \mathbb{Q}^c$, and consequently there exists an $N \in \mathbb{N}$ such that

$$|a_n - a_m| < \frac{\eta}{4}$$

for all $n > N$. Therefore, for all $n > N$ we have

$$\eta - |a_n - a_m| - \frac{\eta}{4} > \frac{\eta}{2} > 0,$$

so that

$$\eta - |a_n - a_m| > \frac{\eta}{4} > 0.$$

Hence
$$\epsilon - |r_m - r| > [(\eta - |a_n - a_m|)] > 0,$$
as required. □

The rational numbers, as discussed in Section 5.5, have the density property: given any two rational numbers a, b for which $a < b$ there is a rational number c such that $a < c < b$. This observation prompts the question of whether the real numbers also have this property. Given any two real numbers r, s for which $r < s$ does there exist a $t \in \mathbb{R}$ such that $r < t < s$? It is plain that this is true for any $r, s \in \mathbb{R}_\mathbb{Q}$, but it is not clear in the case where r or s is in $\mathbb{R}_I$. The next theorem shows that $\mathbb{R}$ has the density property. In fact, there always exists a $t \in \mathbb{R}_\mathbb{Q}$ between r and s.

Theorem 7.12 *Let r and s be two real numbers for which $r < s$. Then there exists a number $t \in \mathbb{R}_\mathbb{Q}$ such that $r < t < s$.*

Proof. Let $r = [(r_n)]$ and $s = [(s_n)]$. Thus $r_n \in \mathbb{Q}$ and $s_n \in \mathbb{Q}$ for each $n \in \mathbb{N}$. Since $s - r > 0$ there is some positive rational number η such that $s - r > [\eta]$. Hence $r + [\eta] < s$, and so there is a natural number N_1 such that
$$r_n + \eta < s_n \tag{7.1}$$
for all $n > N_1$. Since $(r_n), (s_n) \in \mathbb{Q}^c$, there are natural numbers N_2 and N_3 such that
$$|r_n - r_m| < \frac{\eta}{4} \tag{7.2}$$
whenever $m, n > N_2$, and
$$|s_n - s_m| < \frac{\eta}{4} \tag{7.3}$$
whenever $m, n > N_3$. Let $N = \max\{N_1, N_2, N_3\}$. Then inequalities 7.1, 7.2 and 7.3 are all satisfied whenever $m, n > N$. Fix $n > N$, and let $t = r_n + \eta/2$. Evidently, $r_n < t < s_n$ and, for all $m > N$, inequality 7.2 implies
$$t - r_m = t - r_n + r_n - r_m > \frac{\eta}{2} - \frac{\eta}{4} = \frac{\eta}{4} > 0.$$
Similarly, inequalities 7.1 and 7.3 yield
$$s_m - t = s_m - s_n + s_n - t > -\frac{\eta}{4} + \frac{\eta}{2} = \frac{\eta}{4} > 0,$$

since $s_n - t = s_n - r_n - \eta/2 > \eta - \eta/2 = \eta/2$. The above inequalities imply that

$$[(r_n)] < [t] < [(s_n)],$$

and the result follows. □

The theorem above is often stated in the form:

The rational numbers are dense among the real numbers.

By taking $s = r + 1$ we find that for any real r there exists a rational $t > r$. Hence there is also an integer $n > r$.

Theorem 7.11 shows that every Cauchy sequence on $\mathbb{R}_\mathbb{Q}$ is convergent. Theorems 7.11 and 7.12 can be used to establish the remarkable result that *every* Cauchy sequence on $\mathbb{R}$ is convergent. This property is called **completeness**. Completeness is a crucial element in many of the proofs found in analysis. It also signals that $\mathbb{R}$ cannot be extended further by use of Cauchy sequences. The next lemma is the first step towards proving the convergence of every Cauchy sequence on $\mathbb{R}$.

Lemma 7.13 *Let (r_n) be a Cauchy sequence on $\mathbb{R}$ and let (s_n) be a sequence on $\mathbb{R}$ such that, for all $n \in \mathbb{N}$, $r_n < s_n < r_n + 1/n$. Then (s_n) is a Cauchy sequence.*

Proof. Choose some positive real number ϵ. The identity

$$s_n - s_m = s_n - r_n + r_m - s_m + r_n - r_m$$

and the triangle inequality yield

$$|s_n - s_m| \leq |s_n - r_n| + |r_m - s_m| + |r_n - r_m|$$
$$< \frac{1}{n} + \frac{1}{m} + |r_n - r_m|. \qquad (7.4)$$

Now (r_n) is a Cauchy sequence and so there is an $N_1 \in \mathbb{N}$ such that

$$|r_n - r_m| < \frac{\epsilon}{3} \qquad (7.5)$$

for all $m, n > N_1$. Choose $n \in \mathbb{N}$ greater than $\max\{N_1, 3/\epsilon\}$. Then inequality 7.5 is satisfied for all $m, n > N$; moreover, $1/n < \epsilon/3$ and $1/m < \epsilon/3$. Inequality 7.4 thus implies that $|s_n - s_m| < \epsilon$ for all $m, n > N$ and therefore (s_n) is a Cauchy sequence. □

Theorem 7.14 (**Completeness**) *Every Cauchy sequence on $\mathbb{R}$ is convergent.*

Proof. Let (r_n) be a Cauchy sequence on $\mathbb{R}$. By Theorem 7.12, for any $n \in \mathbb{N}$ there is a number $s_n \in \mathbb{R}_\mathbb{Q}$ such that
$$r_n < s_n < r_n + \frac{1}{n}.$$
Lemma 7.13 shows that (s_n) is a Cauchy sequence on $\mathbb{R}_\mathbb{Q}$. Lemma 7.9 and Theorem 7.11 thus imply that $s_n \to s$ for some $s \in \mathbb{R}$. Choose a positive real number ϵ. The identity $r_n - s = r_n - s_n + s_n - s$, the definition of (s_n), and the triangle inequality imply that
$$|r_n - s| < \frac{1}{n} + |s_n - s|.$$
Since $s_n \to s$, there is a natural number N_1 such that $|s_n - s| < \epsilon/2$ for all $n > N_1$. Let $N > \max\{N_1, 2/\epsilon\}$. Then $1/n < \epsilon/2$ and $|s_n - s| < \epsilon/2$ for all $n > N$. Therefore, $|r_n - s| < \epsilon$ for all $n > N$, so that $r_n \to s$ as $n \to \infty$. $\square$

Corollary 7.15 (**The Cauchy Criterion**) *A sequence on $\mathbb{R}$ converges if and only if it is a Cauchy sequence.*

Proof. Theorem 7.14 shows that every Cauchy sequence on $\mathbb{R}$ is convergent. The fact that every convergent sequence on $\mathbb{R}$ is Cauchy can be established using the same arguments *mutatis mutandis* as in the proof of Theorem 6.4. $\square$

7.3 General Results Concerning Sequences

Results for convergent sequences on $\mathbb{Q}$ have straightforward analogues for sequences on $\mathbb{R}$. We present some of these analogues in this section. We omit proofs that are formally identical to those given for sequences on $\mathbb{Q}$.

Theorem 7.16 *If a sequence on $\mathbb{R}$ converges, then the limit of the sequence is unique.*

Proof. See Theorem 6.1. $\square$

Theorem 7.17 *Let (a_n) be a sequence on $\mathbb{R}$ and suppose that $a_n \to a$ as $n \to \infty$. Then every subsequence (a_{n_k}) of (a_n) is convergent and $a_{n_k} \to a$ as $n \to \infty$.*

Proof. See Theorem 6.2. $\square$

Theorem 7.18 *Any convergent sequence on $\mathbb{R}$ is bounded.*

Proof. See Theorem 6.3. $\square$

Theorem 7.19 *Let (a_n) be a convergent sequence on $\mathbb{R}$, let $M, m \in \mathbb{R}$ and let $n \in \mathbb{N}$. If $a_n \leq M$ for all $n > N$ then $\lim_{n \to \infty} a_n \leq M$; if $a_n \geq m$ for all $n > N$ then $\lim_{n \to \infty} a_n \geq m$.*

Proof. Let $a = \lim_{n \to \infty} a_n$ and suppose that $a_n \leq M$ for all $n > N$. Suppose also that $a > M$. Then $a_n < a$ for all $n > N$, and $a - M > 0$. Since $a_n \to a$ as $n \to \infty$ there is an N_1 such that $|a_n - a| < a - M$ for all $n > N_1$. But

$$|a_n - a| = a - a_n \geq a - M$$

for all $n > N$. This contradiction shows that $a \leq M$. The statement concerning a sequence bounded below can be proved in the same way. □

Theorem 7.20 *If (a_n) and (b_n) are sequences on $\mathbb{R}$ and $a_n \to a$, $b_n \to b$ as $n \to \infty$, then*

(a) $\lim_{n \to \infty} (a_n + b_n) = a + b$;
(b) $\lim_{n \to \infty} a_n b_n = ab$;
(c) $\lim_{n \to \infty} a_n / b_n = a/b$, *provided $b \neq 0$.*

Proof. (a) Choose $\epsilon > 0$. For all $n \in \mathbb{N}$

$$|a_n + b_n - (a + b)| = |a_n - a + b_n - b|$$
$$\leq |a_n - a| + |b_n - b|.$$

Since $a_n \to a$ as $n \to \infty$, there is an N_1 such that $|a_n - a| < \epsilon/2$ for all $n > N_1$. Similarly, since $b_n \to b$ as $n \to \infty$, there is an N_2 such that $|b_n - b| < \epsilon/2$ for all $n > N_2$. Let $N = \max\{N_1, N_2\}$. Then $|a_n - a| < \epsilon/2$ and $|b_n - b| < \epsilon/2$ for all $n > N$ and hence

$$|a_n + b_n - (a + b)| \leq |a_n - a| + |b_n - b|$$
$$< \frac{\epsilon}{2} + \frac{\epsilon}{2}$$
$$= \epsilon$$

for all $n > N$. Thus, $a_n + b_n \to a + b$ as $n \to \infty$.

(b) Choose $\epsilon > 0$. For all $n \in \mathbb{N}$

$$|a_n b_n - ab| = |a_n b_n - a_n b + a_n b - ab|$$
$$\leq |a_n||b_n - b| + |b||a_n - a|.$$

Since (a_n) is convergent, Theorem 7.18 implies that it is bounded. Therefore there is an $M > 0$ such that $|a_n| < M$ for all $n \in \mathbb{N}$. Since $b_n \to b$ as

$n \to \infty$, there is an N_1 such that

$$|b_n - b| < \frac{\epsilon}{2M} \tag{7.6}$$

for all $n > N_1$. If $b = 0$, then

$$|a_n b_n - ab| \leq |a_n||b_n - b| < M\frac{\epsilon}{2M} = \frac{\epsilon}{2} < \epsilon$$

for all $n > N_1$. Hence, $a_n b_n \to ab$ as $n \to \infty$. If $b \neq 0$, then there is an N_2 such that

$$|a_n - a| < \frac{\epsilon}{2|b|} \tag{7.7}$$

for all $n > N_2$, since $a_n \to a$ as $n \to \infty$. Let $N = \max\{N_1, N_2\}$. Then inequalities 7.6 and 7.7 are satisfied for all $n > N$; hence,

$$|a_n b_n - ab| \leq |a_n||b_n - b| + |b||a_n - a|$$
$$< M\frac{\epsilon}{2M} + |b|\frac{\epsilon}{2|b|}$$
$$= \epsilon,$$

so that $a_n b_n \to ab$ as $n \to \infty$.

(c) By hypothesis $b_n \to b$ as $n \to \infty$ and $b \neq 0$. Therefore there is an N_1 such that

$$|b_n - b| < \frac{|b|}{2}$$

for all $n > N_1$, and hence

$$b - \frac{|b|}{2} < b_n < b + \frac{|b|}{2}$$

for all $n > N_1$. If $b < 0$ then it follows from the right hand inequality that $b_n < b + |b|/2 = b - b/2 = b/2 = -|b|/2 < 0$. If $b > 0$ then the left hand inequality similarly implies that $b_n > b - |b|/2 = |b|/2 > 0$. In either case $|b_n| > |b|/2 > 0$ for all $n > N_1$. Hence $1/|b_n| < 2/|b|$ for each such n.

For all $n > N_1$ we have

$$\left|\frac{a_n}{b_n} - \frac{a}{b}\right| = \frac{|a_n b - ab_n|}{|b_n b|}$$
$$= \frac{|a_n b - a_n b_n + a_n b_n - ab_n|}{|b_n||b|}$$
$$\leq \frac{2}{|b|^2}(|a_n||b_n - b| + |b_n||a_n - a|).$$

Since (a_n) and (b_n) are convergent, Theorem 7.18 implies that these sequences are bounded. Therefore there exist an $M_1 > 0$ and an $M_2 > 0$ such that $|a_n| < M_1$ and $|b_n| < M_2$ for all $n \in \mathbb{N}$. Since $b_n \to b$ as $n \to \infty$, there is an N_2 such that

$$|b_n - b| < \frac{|b|^2 \epsilon}{4M_1} \tag{7.8}$$

for all $n > N_2$. Similarly, since $a_n \to a$ as $n \to \infty$, there is an N_3 such that

$$|a_n - a| < \frac{|b|^2 \epsilon}{4M_2} \tag{7.9}$$

for all $n > N_3$. Let $N = \max\{N_1, N_2, N_3\}$. Then inequalities 7.8 and 7.9 are satisfied for all $n > N$, and

$$\begin{aligned}\left|\frac{a_n}{b_n} - \frac{a}{b}\right| &\leq \frac{2}{|b|^2}\left(|a_n||b_n - b| + |b_n||a_n - a|\right) \\ &< \frac{2}{|b|^2}\left(M_1 \frac{|b|^2 \epsilon}{4M_1} + M_2 \frac{|b|^2 \epsilon}{4M_2}\right) \\ &= \frac{\epsilon}{2} + \frac{\epsilon}{2} \\ &= \epsilon.\end{aligned}$$

Therefore $a_n/b_n \to a/b$ as $n \to \infty$. $\square$

The following result is a special case of (b).

Corollary 7.21 *For all $c \in \mathbb{R}$ we have*

$$\lim_{n \to \infty} c a_n = c \lim_{n \to \infty} a_n.$$

Theorem 7.22 *If (a_n) is a sequence on $\mathbb{R}$ and $a_n \to a$ as $n \to \infty$, then*

$$\lim_{n \to \infty} |a_n| = |a|.$$

Proof. Choose $\epsilon > 0$. Since $a_n \to a$ as $n \to \infty$, there exists N such that $|a_n - a| < \epsilon$ for all $n > N$. Hence $||a_n| - |a|| \leq |a_n - a| < \epsilon$ for all $n > N$, and the result follows. $\square$

7.4 Sets in $\mathbb{R}$

Recall that each natural number n is equal to the set of all natural numbers less than n. Moreover a set X is finite if there is a bijection from X onto some natural number; otherwise X is infinite. For instance, the set $\{1, 2, 3, 4\}$ is finite; the set $\mathbb{N}$ of natural numbers is infinite.

Lemma 7.23 *Let a and b be real numbers such that $a < b$, and let*
$$X = \{x : a < x < b\}.$$
Then X is an infinite set.

Proof. Theorem 7.12 implies that there is a $t \in \mathbb{R}$ such that $a < t < b$ and therefore X is not empty. Suppose that X is a finite set. Then there is a bijection from X onto the natural number $n = |X| + 1$, and we may write $X = \{x_1, x_2, \ldots, x_{n-1}\}$. Let $x_m = \min X$. Thus $x_m \leq x$ for all $x \in X$. Since $x_m \in X$, we have $a < x_m < b$. Theorem 7.12 implies that there is a $t_m \in \mathbb{R}$ such that $a < t_m < x_m < b$ and therefore $t_m \in X$, so that x_m cannot be the least element in X. We conclude that X is infinite. □

Sets such as X in Lemma 7.23 are of particular interest in analysis. For $a, b \in \mathbb{R}$ such that $a \leq b$ we define the following sets:

$$(a, b) = \{x : a < x < b\};$$
$$[a, b] = \{x : a \leq x \leq b\};$$
$$[a, b) = \{x : a \leq x < b\};$$
$$(a, b] = \{x : a < x \leq b\}.$$

The set (a, b) is called an **open interval**; the set $[a, b]$ is called a **closed interval**. The sets $[a, b)$ and $(a, b]$ are called **semi-open intervals**. The **length** of any of these intervals is defined to be $b - a$. Note that all these intervals include (a, b). If $a < b$, then Lemma 7.23 shows that these intervals are infinite sets. If $a = b$, then $[a, a] = \{a\}$ and $(a, a) = [a, a) = (a, a] = \emptyset$.

For any $a \in \mathbb{R}$ we also define the **semi-infinite intervals**:

$$(a, \infty) = \{x : x > a\};$$
$$[a, \infty) = \{x : x \geq a\};$$
$$(-\infty, a) = \{x : x < a\};$$
$$(-\infty, a] = \{x : x \leq a\}.$$

Sets in ℝ

An element x of a set $X \subseteq \mathbb{R}$ is called an **interior point** of X if there is a $\delta > 0$ such that
$$(x - \delta, x + \delta) \subseteq X.$$
Clearly if $X \subseteq Y \subseteq \mathbb{R}$ then any interior point of X is also an interior point of Y. A set X is called **open** if all its elements are interior points of X.

Example 7.1 Let $a < b$. Prove that (a, b) is an open set.

Solution: We need to show that every point in (a, b) is an interior point. Choose $x \in (a, b)$, so that $a < x < b$. Thus if we let
$$\delta = \min\{x - a, b - x\},$$
then $\delta > 0$. Moreover
$$x - \delta \geq x - (x - a) = a$$
and
$$x + \delta \leq x + (b - x) = b.$$
Therefore $(x - \delta, x + \delta) \subseteq (a, b)$ and hence x is an interior point. Since x is an arbitrary element of (a, b), we conclude that all the elements in (a, b) are interior points of (a, b) and hence that (a, b) is open.

△

Example 7.2 Show that, for any $a \in \mathbb{R}$, the set (a, ∞) is open, but the set $[a, \infty)$ is not open.

Solution: We first show that (a, ∞) is open. Choose $x \in (a, \infty)$, so that $x > a$. Thus if we let $\delta = x - a$ then $\delta > 0$ and $x - \delta = a$. Hence $(x - \delta, x + \delta) \subseteq (a, \infty)$, and thus x is an interior point. Since x is arbitrary, we conclude that all the elements in (a, ∞) are interior points of (a, ∞) and hence that (a, ∞) is open.

To show that the set $[a, \infty)$ is not open, consider $a \in [a, \infty)$. For any $\delta > 0$ we have $(a - \delta, a) \neq \emptyset$ and
$$(a - \delta, a) \cap [a, \infty) = \emptyset.$$
Therefore, there is no $\delta > 0$ such that $(a - \delta, a + \delta) \subseteq [a, \infty)$ and hence a is not an interior point of $[a, \infty)$. Consequently, $[a, \infty)$ is not an open set.

△

An element x of a set $X \subseteq \mathbb{R}$ is called an **isolated point** of X if there is a $\delta > 0$ such that
$$(x - \delta, x + \delta) \cap X = \{x\}.$$
Evidently, isolated points of X cannot be interior points of X and therefore any set containing isolated points cannot be open.

A number $x \in \mathbb{R}$ is called a **limit point** of a set $X \subseteq \mathbb{R}$ if, for all $\delta > 0$, the interval $(x - \delta, x + \delta)$ contains at least one element of $X - \{x\}$. Note that x need not be in X to be a limit point of X. Note also that the requirement that at least one element of $X - \{x\}$ be in $(x - \delta, x + \delta)$ precludes isolated points from being limit points. In fact, since there must be an element of $X - \{x\}$ in $(x - \delta, x + \delta)$ for any $\delta > 0$, it is easy to see that $X \cap (x - \delta, x + \delta)$ must be infinite, for if
$$(X - \{x\}) \cap (x - \delta, x + \delta) = \{x_1, x_2, \ldots, x_n\}$$
then there is no element of $X - \{x\}$ in $(x - \gamma, x + \gamma)$ where γ is any positive number less than
$$\min\{|x_i - x| : i \in \{1, 2, \ldots, n\}\}.$$
More formally, we have the following result.

Lemma 7.24 *Let x be a limit point for the set $X \subseteq \mathbb{R}$. Then there exists a sequence (x_n) such that $x_n \in X - \{x\}$ for all $n \in \mathbb{N}$ and $x_n \to x$ as $n \to \infty$.*

Proof. Let x be a limit point of X. Then for all $n \in \mathbb{N}$ there is a point $x_n \in X - \{x\}$ such that
$$x_n \in \left(x - \frac{1}{n}, x + \frac{1}{n}\right).$$
We may thus construct a sequence (x_n) of points in $X - \{x\}$ such that $|x_n - x| < 1/n$ for all $n \in \mathbb{N}$. Choose $\epsilon > 0$ and let N be any natural number greater than $1/\epsilon$. Then $1/N < \epsilon$ and for all $n > N$ we have
$$|x_n - x| < \frac{1}{n} < \frac{1}{N} < \epsilon.$$
Hence, $x_n \to x$ as $n \to \infty$. □

Note that any interior point of a set must also be a limit point for that set.

Example 7.3 Let $X = \{1/n : n \in \mathbb{N}\}$. Find the limit points of X.

Solution: For any $n \in \mathbb{N}$,
$$\frac{1}{n} - \frac{1}{n+1} = \frac{1}{n(n+1)} > \frac{1}{2n(n+1)} > 0.$$

Thus if $n > 1$ then
$$\frac{1}{n-1} - \frac{1}{n} > \frac{1}{2n(n-1)} > \frac{1}{2n(n+1)},$$

and so
$$\frac{1}{n+1} < \frac{1}{n} - \frac{1}{2n(n+1)} < \frac{1}{n} < \frac{1}{n} + \frac{1}{2n(n+1)} < \frac{1}{n-1};$$

hence
$$\left(\frac{1}{n} - \frac{1}{2n(n+1)}, \frac{1}{n} + \frac{1}{2n(n+1)}\right) \cap X = \left\{\frac{1}{n}\right\}$$

for all $n > 1$. As 1 is certainly an isolated point of X, we infer that every point of X is therefore an isolated point. But $1/n \to 0$ as $n \to \infty$, so that for any $\delta > 0$ there is an N such that $0 < 1/n < \delta$ for all $n > N$ (cf. Example 6.1). Therefore, 0 is a limit point of X.

On the other hand, suppose that $r \neq 0$ is a limit point of X. Then $(r - \delta, r + \delta) \cap X$ must be infinite for each $\delta > 0$. Taking $\delta = |r|/2$ we find that this conclusion contradicts the convergence of the sequence $(1/n)$ to 0. It follows that 0 is the only limit point of X.

△

Example 7.4 Find all the limit points for (a, b) where $a < b$.

Solution: The set (a, b) is open (Example 7.1) and hence every point of (a, b) is an interior point. Every interior point must also be a limit point and hence all the points of (a, b) are limit points of (a, b). The numbers a and b are also limit points for this set. To show that a is a limit point, choose any $\delta > 0$ and let $\mu = \min\{\delta, b-a\}$. Then $(a-\delta, a+\delta) \cap (a,b) = (a, a+\mu)$. Since $\mu > 0$, $(a, a+\mu)$ is an infinite set by Lemma 7.23 and hence a is a limit point of (a, b). A similar argument can be used to show that b is also a limit point of (a, b).

It may be that (a, b) has other limit points not in the interval $[a, b]$. Suppose that $x < a$, and let $\delta = a - x > 0$. It follows that $x + \delta = a$, so that $(x - \delta, x + \delta) \cap (a, b) = \emptyset$ and therefore x cannot be a limit point of (a, b). A similar argument shows that (a, b) has no limit points greater than b. The set of all limit points of (a, b) is therefore the closed interval $[a, b]$.

△

A set $X \subseteq \mathbb{R}$ is **closed** if X contains all its limit points. Since finite sets cannot have limit points, they are vacuously closed. Infinite sets of isolated points, such as the set $\mathbb{N}$ of all natural numbers, are also closed. Example 7.3, however, shows that there is an infinite set X of isolated points such that X has a limit point. From Example 7.4 we see that (a, b) is not a closed set but that $[a, b]$ is closed.

Example 7.5 Show that $[a, \infty)$ is a closed set for any $a \in \mathbb{R}$.

Solution: Example 7.2 shows that (a, ∞) is open and so any point in (a, ∞) is a limit point in $[a, \infty)$. The arguments of Example 7.4 show that a is also a limit point for $[a, \infty)$ and that no limit point is less than a. Hence the set of limit points for $[a, \infty)$ is $[a, \infty)$, as required.

△

The **closure** $\bar{X}$ of a set $X \subseteq \mathbb{R}$ is the union of X with its set of limit points. The closure of the set X of Example 7.3, for instance, is $X \cup \{0\}$. From Example 7.4 we see that the closure of the open interval (a, b) for $a < b$ is $[a, b]$. It can be easily shown that the closure of a set is closed, though we do not need this result.

Note that there are sets that are neither open nor closed. For example, the set $[a, b)$ where $a < b$ cannot be open since $a \in [a, b)$ and a is not an interior point, and it cannot be closed since b is a limit point for $[a, b)$ and $b \notin [a, b)$. By contrast, there are sets that are both open and closed. An example is the set of all real numbers. The empty set also has this property vacuously. Open sets and closed sets are related through complementation, not negation. We leave it to the reader to prove that a set $X \subseteq \mathbb{R}$ is closed if and only if $\mathbb{R} - X$ is open. However this result is not needed in the sequel.

7.5 Bounded Sets in $\mathbb{R}$

A set $X \subseteq \mathbb{R}$ is **bounded above** if there exists a real number Λ such that $x < \Lambda$ for all $x \in X$. The number Λ is called an **upper bound** of X. Note that it is not necessary for Λ to be a member of X and that Λ is not unique. If Λ is in X then it is clear that there are no other members of X that are also upper bounds. In this case Λ is called the **maximum element** of X. Thus the maximum element, if it exists, is unique. Similar definitions can be framed for a set **bounded below**, a **lower bound**, and a **minimum element**.

Example 7.6 Let X be the set of all negative real numbers. An upper bound of X is 0. Any positive real number is also an upper bound of X. The set X does not have a maximum element: if $x \in X$ were a maximum element then Theorem 7.12 would show that there is a $t \in \mathbb{R}$ such that $x < t < 0$ and consequently $t \in X$ and $t > x$. Evidently X is not bounded below.

△

Example 7.7 Let $X = \{1, -3, 4/7, -1/2\}$. An upper bound of X is 1, but any number greater than 1 is also an upper bound. Here 1 is also the maximum element because $1 \in X$. Similarly, -3 is a lower bound of X as is any number less than -3. The number -3 is also the minimum element of X.

△

In Example 7.6 the number 0 is an upper bound of X but not a maximum element, yet it is still distinguished from the other upper bounds in that there are no upper bounds less than 0 for this set. In other words, if Λ is an upper bound of X then $\Lambda \geq 0$. The set of Example 7.7 has a maximum element. It is also clear that for this set there are no upper bounds less than this maximum element. Upper bounds such as 0 for Example 7.6 and 1 for Example 7.7 are of particular interest in analysis and thus given a special name. For a subset X of $\mathbb{R}$, an element $\Lambda \in \mathbb{R}$ is the **least upper bound**, lub X, of X if Λ is an upper bound of X and $\Gamma \geq \Lambda$ for every upper bound Γ. It is clear that the least upper bound of X is unique and may or may not be an member of X. In the same spirit, a greatest lower bound can also be defined. For a set $X \subseteq \mathbb{R}$, the element $\lambda \in \mathbb{R}$ is the **greatest lower bound**, glb X, of X if $\gamma \leq \lambda$ for every lower bound γ. As with the least upper bound, λ is unique and may or may not be a member of X. If X is finite then lub $X = \max X$ and glb $X = \min X$.

The main result of this section is that every nonempty set $X \subseteq \mathbb{R}$ that is bounded above has a least upper bound. The next two lemmas are useful in the proof of this result.

Lemma 7.25 *For any $n \in \mathbb{N} \cup \{0\}$,*

$$\sum_{k=0}^{n}(1/2)^k < 2.$$

Proof. By Corollary 4.30 we have

$$\sum_{k=0}^{n}\left(\frac{1}{2}\right)^k = \frac{\left(\frac{1}{2}\right)^{n+1} - 1}{\frac{1}{2} - 1}$$
$$= 2\left(1 - \frac{1}{2^{n+1}}\right)$$
$$< 2.$$

$\square$

In preparation for the next lemma, we note by induction that $2^n > n$ for any $n \in \mathbb{N}$. Indeed, $2^1 = 2 > 1$, and if $2^n > n$ for some $n \in \mathbb{N}$ then $2^{n+1} = 2 \cdot 2^n > 2n \geq n + 1$, since $n \geq 1$ (cf. problem 6, Chapter 3). It follows that

$$\frac{1}{2^n} < \frac{1}{n}.$$

Lemma 7.26 *For any positive real numbers s, ϵ there exists an $N \in \mathbb{N}$ such that*

$$\frac{s}{2^n} < \epsilon$$

for all $n > N$.

Proof. Choose any positive real number ϵ. Lemma 7.7 implies that for any positive real number s there is a positive $\eta \in \mathbb{Q}$ such that $\eta < \epsilon/s$. Moreover, since η is a positive rational number it can be written as p/N where p and N are natural numbers. Hence $\eta \geq 1/N$ and consequently $1/N < \epsilon/s$, so that $s/N < \epsilon$. Therefore

$$\frac{s}{2^n} < \frac{s}{2^N} < \frac{s}{N} < \epsilon$$

for any $n > N$ since $s > 0$. $\square$

Theorem 7.27 *Every nonempty subset of $\mathbb{R}$ that is bounded above has a least upper bound.*

Proof. Let $X \subseteq \mathbb{R}$ be nonempty and bounded above. Let x_1 be an element of X and Λ_1 an upper bound of X. Let $t_1 = x_1 - 1$ and $C_1 = (\Lambda_1 + t_1)/2$. Then $t_1 < x_1 \leq \Lambda_1$ and $t_1 < C_1 < \Lambda_1$, and there are two possibilities:

(a) C_1 is an upper bound for X, or
(b) C_1 is not an upper bound for X.

If (a) is true, let $t_2 = t_1$ and $\Lambda_2 = C_1$, so that Λ_2 is an upper bound. Then

$$\Lambda_2 - t_2 = \frac{\Lambda_1 - t_1}{2} > 0 \tag{7.10}$$

and

$$0 \leq \Lambda_1 - \Lambda_2 \leq \frac{\Lambda_1 - t_1}{2}. \tag{7.11}$$

Since Λ_2 is an upper bound, $x \leq \Lambda_2$ for all $x \in X$. Putting $x_2 = x_1$, we therefore have $t_2 < x_2 \leq \Lambda_2$. If (b) is true then let $t_2 = C_1$ and $\Lambda_2 = \Lambda_1$. Relations 7.10 and 7.11 are still valid and there exists an $x_2 \in X$ such that $t_2 < x_2 \leq \Lambda_2$ since t_2 is not an upper bound. In both cases Λ_2 is an upper bound.

For each natural number $n > 1$ we now establish the existence of numbers x_n, t_n, Λ_n such that

$$0 \leq \Lambda_{n-1} - \Lambda_n \leq \frac{\Lambda_1 - t_1}{2^{n-1}} = \Lambda_n - t_n,$$

where $t_n < x_n \leq \Lambda_n$, $x_n \in X$ and Λ_n is an upper bound of X. We already have the existence of x_2, t_2 and Λ_2, and so we may assume that of x_{n-1}, t_{n-1} and Λ_{n-1}, where $n > 2$. Let

$$C_{n-1} = \frac{\Lambda_{n-1} + t_{n-1}}{2}.$$

If C_{n-1} is an upper bound of X, then let $t_n = t_{n-1}$ and $\Lambda_n = C_{n-1}$; otherwise put $t_n = C_{n-1}$ and $\Lambda_n = \Lambda_{n-1}$. In both cases

$$\Lambda_n - t_n = \frac{\Lambda_{n-1} - t_{n-1}}{2} = \frac{\Lambda_1 - t_1}{2^{n-1}}$$

and

$$0 \leq \Lambda_{n-1} - \Lambda_n$$
$$\leq \Lambda_{n-1} - t_{n-1} + t_n - \Lambda_n$$
$$= \frac{\Lambda_1 - t_1}{2^{n-2}} - \frac{\Lambda_1 - t_1}{2^{n-1}}$$
$$= \frac{2(\Lambda_1 - t_1) - (\Lambda_1 - t_1)}{2^{n-1}}$$
$$= \frac{\Lambda_1 - t_1}{2^{n-1}}.$$

Moreover Λ_n is an upper bound of X but t_n is not, and so there exists $x_n \in X$ such that $t_n < x_n \leq \Lambda_n$. The existence of the required x_n, t_n, Λ_n has now been established.

It will now be shown that (Λ_n) is a Cauchy sequence on $\mathbb{R}$. Choose some positive real number ϵ and let $\alpha = \Lambda_1 - t_1$. Let m, n be positive integers such that $n > m$. The difference $\Lambda_m - \Lambda_n$ can be bounded as follows: by the telescoping property,

$$|\Lambda_m - \Lambda_n| = \Lambda_m - \Lambda_n$$
$$= \sum_{k=m}^{n-1} (\Lambda_k - \Lambda_{k+1})$$
$$\leq \alpha \sum_{k=m}^{n-1} \frac{1}{2^k}$$
$$= \frac{\alpha}{2^m} \sum_{k=m}^{n-1} \frac{1}{2^{k-m}}$$
$$= \frac{\alpha}{2^m} \sum_{k=0}^{n-m-1} \frac{1}{2^k}$$
$$< \frac{\alpha}{2^{m-1}},$$

where the last inequality follows from Lemma 7.25. Since $2\alpha > 0$, Lemma 7.26 implies that there is an $N \in \mathbb{N}$ such that

$$\frac{\alpha}{2^{m-1}} < \epsilon$$

for all $m > N$. Therefore $|\Lambda_m - \Lambda_n| < \epsilon$ for all $n, m > N$ and (Λ_n) is thus a Cauchy sequence.

Evidently (Λ_n) is a sequence of upper bounds of X, and Theorem 7.14 implies that $\Lambda_n \to \Lambda$ as $n \to \infty$ for some $\Lambda \in \mathbb{R}$. It will now be shown that Λ is an upper bound of X. Suppose that Λ is not an upper bound of X. Then there is some $x \in X$ such that $\Lambda < x$. Since $x - \Lambda$ is positive and $\Lambda_n \to \Lambda$, there must be some $M \in \mathbb{N}$ such that

$$|\Lambda_n - \Lambda| < x - \Lambda$$

for all $n > M$. Hence $\Lambda_n - \Lambda < x - \Lambda$, so that $\Lambda_n < x$ for each $n > M$. This result contradicts the fact that Λ_n is an upper bound of X; consequently, Λ is an upper bound of X.

The theorem will be established if it can be shown that any other upper bound of X must be greater than Λ. Let $\Gamma < \Lambda$. We must show that Γ cannot be an upper bound of X. The identity $\Lambda - x_n = \Lambda - \Lambda_n + \Lambda_n - x_n$, the inequality $t_n < x_n \leq \Lambda_n$ and the triangle inequality yield the inequality

$$|\Lambda - x_n| < |\Lambda - \Lambda_n| + |\Lambda_n - t_n|$$

for all $n \in \mathbb{N}$. Let $\epsilon = \Lambda - \Gamma > 0$. Since $\Lambda_n \to \Lambda$, there is an $N_1 \in \mathbb{N}$ such that, for any $n > N_1$,

$$|\Lambda - \Lambda_n| < \frac{\epsilon}{2};$$

moreover, since $2\alpha > 0$ and

$$|\Lambda_n - t_n| = \Lambda_n - t_n = \frac{\alpha}{2^{n-1}},$$

Lemma 7.26 implies that there is some $N_2 \in \mathbb{N}$ such that

$$|\Lambda_n - t_n| < \frac{\epsilon}{2}$$

for all $n > N_2$. Therefore,

$$|\Lambda - x_n| < \epsilon$$

for all $n > N = \max\{N_1, N_2\}$, and consequently

$$-\epsilon < \Lambda - x_n < \epsilon = \Lambda - \Gamma$$

for each such n. Therefore $x_n > \Gamma$ for all $n > N$. But $x_n \in X$ and therefore Γ cannot be an upper bound. We conclude that $\Lambda = \text{lub } X$. □

Corollary 7.28 *Every nonempty subset of $\mathbb{R}$ that is bounded below has a greatest lower bound.*

Proof. Let X be a nonempty set bounded below and let
$$Y = \{-x : x \in X\}.$$
Since X is bounded below, Y must be bounded above. The set Y is nonempty since X is nonempty and hence Y has a least upper bound by Theorem 7.27. Let $\Lambda = \text{lub } Y$. Then $\Lambda \geq -x$ for all $x \in X$, and if Γ is any other upper bound of Y then $\Gamma \geq \Lambda$. Therefore $x \geq -\Lambda$ for all $x \in X$, and $-\Lambda$ is therefore a lower bound of X. On the other hand, let γ be any lower bound of X. Then $-\gamma$ is an upper bound of Y. Thus $\Lambda \leq -\gamma$, and so $\gamma \leq -\Lambda$. Hence, $-\Lambda$ is the greatest lower bound of X. □

As noted earlier, if a set X has a least upper bound then lub X need not be in X. If lub $X \notin X$, however, the next result shows that lub X must be a limit point of X. A similar statement holds for sets with greatest lower bounds.

Theorem 7.29 *Let $X \subseteq \mathbb{R}$ be a nonempty set.*

(a) *If X has a greatest lower bound then either* glb $X \in X$ *or* glb X *is a limit point of X.*
(b) *If X has a least upper bound then either* lub $X \in X$ *or* lub X *is a limit point of X.*

Proof. Let X be a set with greatest lower bound λ. Suppose that $\lambda \notin X$ and that λ is not a limit point of X. Then there is a $\delta > 0$ such that $(\lambda - \delta, \lambda + \delta) \cap X = \emptyset$ and hence $\lambda + \delta \leq x$ for all $x \in X$. Thus $\lambda + \delta$ is a lower bound of X, and so λ cannot be the greatest lower bound. This contradiction shows that λ must be a limit point if $\lambda \notin X$.

The proof of (b) is similar. □

Note that a greatest lower bound or a least upper bound can be both an element of X and a limit point of X. For example, if $X = [a, b)$, then glb $X = a \in X$, and a is also a limit point for X.

A set is called **compact** it it is closed and bounded. For example, the closed interval $[a, b]$ is a compact set.

Theorem 7.30 *Let $X \subseteq \mathbb{R}$ be a nonempty compact set. Then X has a greatest lower bound and a least upper bound. Moreover,* glb $X \in X$ *and* lub $X \in X$.

Proof. Since X is compact it is bounded above and below. The set X is nonempty and hence Corollary 7.28 and Theorem 7.27 imply that X has a greatest lower bound and a least upper bound. Theorem 7.29 shows that either glb $X \in X$ or glb X is a limit point of X. But X is closed and therefore contains all its limit points. Therefore glb $X \in X$. It can be shown similarly that lub $X \in X$. □

7.6 Monotonic Sequences

The definition of convergence can be used to prove that a specific sequence on $\mathbb{R}$ converges or diverges, but in the case of convergence it is sometimes required to find the limit. The limit is not always obvious. Consider for example the sequence (b_n) on $\mathbb{R}$ defined by

$$b_n = \left(1 + \frac{1}{n}\right)^n$$

for all $n \in \mathbb{N}$. We show in Example 7.13 that this sequence is convergent, but it is not obvious what the limit is. The Cauchy criterion (Corollary 7.15) is a useful result for convergence questions because it does not require a candidate for a limit. Nonetheless, for a sequence such as (b_n) this result is not particularly convenient to use. Fortunately, there are results for certain types of sequences, called monotonic sequences, that circumvent the problem of finding a candidate for the limit. In this section we discuss monotonic sequences and present a basic result linking these sequences with convergence. We also present another result applicable to general sequences.

Let (a_n) be a sequence on $\mathbb{R}$. If $a_{n+1} \geq a_n$ for all $n \in \mathbb{N}$ then the sequence is **non-decreasing**; if $a_{n+1} \leq a_n$ for all $n \in \mathbb{N}$ then the sequence is **non-increasing**. A sequence is said to be **monotonic** if it is either non-decreasing or non-increasing.

Theorem 7.31 *Let (a_n) be a sequence on $\mathbb{R}$.*

(a) *If (a_n) is a non-decreasing sequence and bounded above by a number M, then (a_n) converges to a limit no greater than M.*
(b) *If (a_n) is a non-increasing sequence and bounded below by a number m, then (a_n) converges to a limit no less than m.*

Proof. (a) Suppose that (a_n) is a non-decreasing sequence and that $a_n \leq M$ for all $n \in \mathbb{N}$. Then the set $A = \{a_n : n \in \mathbb{N}\}$ is nonempty and bounded above by M. Theorem 7.27 implies that A must have a least upper bound, a. Evidently, $a_n \leq a \leq M$ for all $n \in \mathbb{N}$. We show that $a_n \to a$ as $n \to \infty$.

Choose $\epsilon > 0$. We must find an N such that
$$|a_n - a| < \epsilon$$
for all $n > N$. Certainly $a_n < a + \epsilon$ for all $n \in \mathbb{N}$. Since $a - \epsilon < a$ and $a = \text{lub } A$, $a - \epsilon$ cannot be an upper bound of A. Thus there must be an $N \in \mathbb{N}$ such that $a - \epsilon < a_N$. But (a_n) is non-decreasing, so that for all $n > N$ we have
$$a_n \geq a_N > a - \epsilon.$$
Therefore, for all $n > N$ it follows that $a - \epsilon < a_n < a + \epsilon$, and so
$$|a_n - a| < \epsilon.$$
Thus, $a_n \to a$ as $n \to \infty$.

(b) Suppose that (a_n) is a non-increasing sequence and that $a_n \geq m$ for all $n \in \mathbb{N}$. Then the sequence (c_n) defined by $c_n = -a_n$ for all $n \in \mathbb{N}$ is bounded above by $-m$ and non-decreasing. Therefore statement (a) of this theorem shows that (c_n) converges to a number no greater than $-m$. Hence (a_n) converges to a number no less than m. □

Example 7.8 Let $q \in \mathbb{R}$ be a number such that $0 < q < 1$ and define the sequence (a_n) by $a_n = q^n$ for all $n \in \mathbb{N}$. Prove that $a_n \to 0$ as $n \to \infty$.

Solution: Since $0 < q < 1$, we have $q^n > q^{n+1} > 0$ for all n. Therefore (a_n) is a non-increasing sequence bounded below by 0. Theorem 7.31 shows that (a_n) is convergent. Let $a = \lim_{n \to \infty} a_n$. Since $a_{n+1} = qa_n$ for all n and $a_{n+1} \to a$ as $n \to \infty$ we have $a = qa$. Now $q \neq 1$ and therefore $a = 0$. △

Example 7.9 Consider the sequence (a_n) where
$$a_n = \sum_{k=0}^{n} \frac{1}{k!}$$
for all $n \geq 0$. This sequence is certainly non-decreasing. Moreover for all $n > 1$ we have
$$2 < a_n \leq 1 + \sum_{k=1}^{n} \frac{1}{2^{k-1}} = 1 + \sum_{k=0}^{n-1} \frac{1}{2^k} < 3$$

by Lemma 7.25. Hence (a_n) is convergent. The limit of this sequence turns out to be a prominent number in analysis. It is customary to denote it by e. Clearly $e \geq a_2 > 2$. Moreover for all $n \geq 3$ we see that

$$a_n \leq 1 + 1 + \frac{1}{2} + \frac{1}{6} + \sum_{k=3}^{n-1} \frac{1}{2^k}$$

$$= 1 + \sum_{k=0}^{n-1} \frac{1}{2^k} - \frac{1}{4} + \frac{1}{6}$$

$$< 3 - \frac{1}{12}$$

$$= \frac{35}{12}.$$

Hence

$$e \leq \frac{35}{12} < 3.$$

We show that e is irrational. Suppose that $e = m/n$ for some $m \in \mathbb{N}$ and $n \in \mathbb{N}$. Since $2 < e < 3$, it follows that $n > 1$. For all $k \geq 1$ we must have

$$a_{n+k} - a_n = \sum_{j=1}^{k} \frac{1}{(n+j)!}$$

$$\leq \frac{1}{(n+1)!} \sum_{j=1}^{k} \frac{1}{(n+1)^{j-1}}$$

$$= \frac{1}{(n+1)!} \sum_{j=0}^{k-1} \frac{1}{(n+1)^j}$$

$$= \frac{1}{(n+1)!} \left(\frac{\frac{1}{(n+1)^k} - 1}{\frac{1}{n+1} - 1} \right)$$

$$= \frac{1}{(n+1)!} \left(\frac{1 - \frac{1}{(n+1)^k}}{\frac{n}{n+1}} \right)$$

$$< \frac{1}{n!n},$$

where Corollary 4.30 was used to evaluate the summation. We deduce that

$$\frac{n!}{(n+1)!} \leq n!(a_{n+k} - a_n) < \frac{1}{n}.$$

Hence
$$0 < \frac{n!}{(n+1)!} \le \lim_{k\to\infty} n!(a_{n+k} - a_n) \le \frac{1}{n} < 1. \tag{7.12}$$

But
$$\lim_{k\to\infty} n! a_{n+k} = n! e = (n-1)! m,$$

which is an integer, as is $n! a_n$. Hence $\lim_{k\to\infty} n!(a_{n+k} - a_n)$ is an integer. This contradiction to 7.12 shows that e is irrational. △

Example 7.10 Let (a_n) be the sequence on $\mathbb{R}$ defined by $a_0 = 1$, and
$$a_n = \frac{1 + a_{n-1}}{2 + a_{n-1}},$$
for all $n \ge 1$. Prove that (a_n) is convergent.

Solution: A simple induction argument shows that $a_n > 0$ for all n. We now use an induction argument to show that (a_n) is non-increasing. We have $a_0 = 1 > 2/3 = a_1$. Suppose that $a_n \le a_{n-1}$. For all $n \ge 1$ we have
$$\frac{1 + a_{n-1}}{2 + a_{n-1}} = 1 - \frac{1}{2 + a_{n-1}},$$
and as $a_n > 0$ it follows that
$$a_{n+1} - a_n = \frac{1 + a_n}{2 + a_n} - \frac{1 + a_{n-1}}{2 + a_{n-1}}$$
$$= \frac{1}{2 + a_{n-1}} - \frac{1}{2 + a_n}$$
$$= \frac{a_n - a_{n-1}}{(2 + a_{n-1})(2 + a_n)}.$$

Since $a_n \le a_{n-1}$, we deduce that $a_{n+1} - a_n \le 0$. Thus $a_{n+1} \le a_n$ for all n. Hence (a_n) is a non-increasing sequence bounded below by 0. Therefore, by Theorem 7.31, (a_n) converges. It is possible to find the limit for this sequence. We revisit this example in Section 7.8. △

Another important result that circumvents the problem of finding a candidate for a limit is the so-called sandwich theorem.

Theorem 7.32 (**Sandwich Theorem**) *Let (a_n), (b_n) and (c_n) be sequences on $\mathbb{R}$ such that $a_n \to b$ and $c_n \to b$ as $n \to \infty$. Suppose that*
$$a_n \leq b_n \leq c_n$$
for all $n \in \mathbb{N}$. Then $b_n \to b$ as $n \to \infty$.

Proof. Choose $\epsilon > 0$. Since $a_n \to b$ as $n \to \infty$ there is an N_1 such that $|a_n - b| < \epsilon$ for all $n > N_1$. Similarly, there is an N_2 such that $|c_n - b| < \epsilon$ for all $n > N_2$. Let $N = \max\{N_1, N_2\}$. Then $|a_n - b| < \epsilon$ and $|c_n - b| < \epsilon$ for all $n > N$. Therefore $b - \epsilon < a_n$ and $c_n < b + \epsilon$ for all $n > N$. Consequently,
$$b - \epsilon < a_n \leq b_n \leq c_n < b + \epsilon$$
for all $n > N$. Hence $|b_n - b| < \epsilon$ for all $n > N$, and we conclude that $b_n \to b$ as $n \to \infty$. $\square$

Example 7.11 Let $b_n = 1/2^n$ for all $n \in \mathbb{N}$. Prove that $b_n \to 0$ as $n \to \infty$.

Solution: It is clear that $b_n > 0$ for all $n \in \mathbb{N}$. We also know from the remark before the statement of Lemma 7.26 that $1/2^n < 1/n$. Let $(a_n) = (0)$ and $(c_n) = (1/n)$. Then $a_n \leq b_n \leq c_n$ for all n, and $a_n \to 0$ and $c_n \to 0$ as $n \to \infty$. Therefore, by Theorem 7.32, $b_n \to 0$ as $n \to \infty$. $\triangle$

Example 7.12 Let (b_n) be the sequence defined by
$$b_n = \frac{1}{n^2 + 1} + \frac{1}{n^2 + 2} + \cdots + \frac{1}{n^2 + n}$$
for all $n \geq 1$. Show that $b_n \to 0$ as $n \to \infty$.

Solution: Note that
$$\frac{1}{n^2 + 1} > \frac{1}{n^2 + j}$$
for any $j > 1$. Therefore,
$$b_n \leq \frac{n}{n^2 + 1} < \frac{n}{n^2} = \frac{1}{n}.$$
Note also that
$$\frac{1}{n^2 + j} > \frac{1}{n^2 + n}$$

for any $j \in \mathbb{N}$ such that $j < n$, so that
$$b_n \geq \frac{n}{n^2+n} = \frac{1}{n+1}.$$
Let $a_n = 0$ and $c_n = 1/n$. Then $a_n < b_n < c_n$ for all $n \geq 1$. Since $a_n \to 0$ and $c_n \to 0$ as $n \to \infty$ we have by Theorem 7.32 that $b_n \to 0$ as $n \to \infty$. △

Before presenting our next example, we give a lemma.

Lemma 7.33 *Let (c_k) be a sequence of real numbers such that $c_j c_k \geq 0$ and $c_k > -1$ for all $j, k \in \mathbb{N}$. For each $n \in \mathbb{N}$ let*
$$P_n = \prod_{k=1}^{n}(1+c_k).$$

Then
$$P_n \geq 1 + \sum_{k=1}^{n} c_k$$

for all $n \in \mathbb{N}$.

Proof. Both sides are equal to $1+c_1$ if $n=1$. If the inequality holds for some $n \in \mathbb{N}$, then since $1+c_j > 0$ for all $j \in \mathbb{N}$ we have
$$\begin{aligned}
P_{n+1} &= (1+c_{n+1})P_n \\
&\geq (1+c_{n+1})\left(1+\sum_{k=1}^{n} c_k\right) \\
&= 1 + \sum_{k=1}^{n+1} c_k + c_{n+1}\sum_{k=1}^{n} c_k \\
&\geq 1 + \sum_{k=1}^{n+1} c_k,
\end{aligned}$$

and the proof is complete by induction. □

Example 7.13 Let (b_n) be the sequence defined by
$$b_n = \left(1 + \frac{1}{n}\right)^n$$

for all $n \geq 1$. We show that $b_n \to e$ as $n \to \infty$. We will use the sequence (a_n) defined in Example 7.9 and the sandwich theorem to establish this result.

For all $n \geq 1$ the binomial theorem (Theorem 5.11) gives

$$b_n = \sum_{k=0}^{n} \binom{n}{k} \frac{1}{n^k}$$

$$= 1 + \sum_{k=1}^{n} \left(\frac{1}{k! n^k} \prod_{j=0}^{k-1} (n-j) \right)$$

$$= 1 + \sum_{k=1}^{n} \left(\frac{1}{k!} \prod_{j=0}^{k-1} \left(1 - \frac{j}{n}\right) \right).$$

Let $d_{n0} = 1$, and for all $k \geq 1$ let

$$d_{nk} = \prod_{j=0}^{k-1} \left(1 - \frac{j}{n}\right).$$

Then

$$b_n = \sum_{k=0}^{n} \frac{d_{nk}}{k!}.$$

Now, $d_{nk} \leq 1$ for all $k \geq 0$ and $n \geq 1$; hence,

$$b_n \leq \sum_{k=0}^{n} \frac{1}{k!} = a_n.$$

To get a lower bound for b_n note that $0 < j/n < 1$ for all j and n such that $0 < j < n$. Therefore we can apply Lemma 7.33 with $c_j = -j/n$ to assert that, for $k \geq 2$,

$$d_{nk} = \prod_{j=1}^{k-1} \left(1 - \frac{j}{n}\right)$$

$$\geq 1 - \frac{1}{n} \sum_{j=1}^{k-1} j$$

$$= 1 - \frac{1}{n} \cdot \frac{k(k-1)}{2},$$

where we have used the result in Example 4.1 to evaluate the summation. By inspection we see that this result also holds if $k \in \{0, 1\}$. For all $n \geq 2$ we thus have

$$b_n = \sum_{k=0}^{n} \frac{d_{nk}}{k!}$$
$$\geq \sum_{k=0}^{n} \frac{1}{k!} \left(1 - \frac{k(k-1)}{2n}\right)$$
$$= \sum_{k=0}^{n} \frac{1}{k!} - \frac{1}{2n} \sum_{k=2}^{n} \frac{1}{(k-2)!}$$
$$= a_n - \frac{1}{2n} a_{n-2}.$$

In summary, for all $n \geq 2$,

$$a_n - \frac{1}{2n} a_{n-2} \leq b_n \leq a_n.$$

We know from Example 7.9 that $a_n \to e$ as $n \to \infty$. Using Example 6.1 we deduce that $a_n - a_{n-2}/(2n) \to e$ as $n \to \infty$. The sandwich theorem thus implies that $b_n \to e$ as $n \to \infty$.

△

7.7 The Bolzano-Weierstrass Theorem

Theorem 7.18 shows that any convergent sequence is bounded. If a sequence is bounded, however, it need not be convergent. The sequence $((-1)^n)$ of Example 6.3 is bounded but divergent. The range of this sequence is the set $\{-1, 1\}$, which is finite. Suppose we consider a sequence such as $((-1)^n + 1/n)$. This sequence is bounded and its range is an infinite set. The subsequence $((-1)^{2n} + 1/(2n))$ converges to 1 and the subsequence $((-1)^{2n+1} + 1/(2n+1))$ converges to -1, so that by Theorem 7.17 the sequence is divergent. Nonetheless, the sequence $((-1)^n + 1/n)$ is fundamentally different from the sequence $((-1)^n)$ because the range of the former has limit points (cf. Example 7.3), whereas the range of the latter is a finite set and cannot have limit points. These simple observations prompt questions concerning infinite sets, bounded sets and limit points. The Bolzano-Weierstrass Theorem links these properties. Essentially this theorem states that infinite sets that are bounded must have at least one limit point.

Theorem 7.34 (**Nested Intervals**) *Let (I_n) be a sequence of closed intervals such that $I_{n+1} \subseteq I_n$ for each $n \in \mathbb{N}$, and the length of I_n approaches 0 as $n \to \infty$. Then there is a unique real number a such that $a \in I_n$ for all $n \in \mathbb{N}$.*

Proof. Let $I_n = [a_n, b_n]$ for each $n \in \mathbb{N}$. Since $I_{n+1} \subseteq I_n$ for each $n \in \mathbb{N}$, we have $a_n \leq a_{n+1}$ and $b_n \geq b_{n+1}$ for all $n \in \mathbb{N}$. Moreover, $a_1 \leq a_n \leq b_n \leq b_1$ for all n. Hence, (a_n) is a non-decreasing sequence bounded above by b_1 and therefore convergent. Similarly, (b_n) is a non-increasing sequence bounded below by a_1 and therefore convergent. Let $a = \lim_{n \to \infty} a_n$ and $b = \lim_{n \to \infty} b_n$. The proof of Theorem 7.31 shows that a is the least upper bound of the set $A = \{a_n : n \in \mathbb{N}\}$ and that b is the greatest lower bound of the set $B = \{b_n : n \in \mathbb{N}\}$. For any fixed $n \in \mathbb{N}$ we have $a_m \leq b_n$ for all $m > n$, since $I_m \subseteq I_n$. Hence $a \leq b_n$ for each $n \in \mathbb{N}$ by Theorem 7.19, and so $a \leq b$, again by Theorem 7.19. Thus

$$a_n \leq a \leq b \leq b_n$$

for all n, and so $a \in I_n$ for all $n \in \mathbb{N}$. Moreover, by hypothesis $b_n - a_n \to 0$ as $n \to \infty$. Since (a_n) and (b_n) are convergent, Theorem 7.20 implies that

$$\begin{aligned} 0 &= \lim_{n \to \infty} (b_n - a_n) \\ &= \lim_{n \to \infty} b_n - \lim_{n \to \infty} a_n \\ &= b - a; \end{aligned}$$

hence, $a = b$.

Suppose that there is a $c \in \mathbb{R}$ such that $c \in I_n$ for all n. Then $a_n \leq c \leq b_n$ for all n. Now, $a_n \to a$ and $b_n \to a$ as $n \to \infty$, and so Theorem 7.32 implies that $c = a$. We conclude that there is only one point $a \in \mathbb{R}$ such that $a \in I_n$ for all n. □

Theorem 7.35 (**Bolzano – Weierstrass**) *An infinite set $X \subseteq \mathbb{R}$ that is bounded has at least one limit point.*

Proof. Since X is bounded there exist $a, b \in \mathbb{R}$ such that $X \subseteq [a, b]$. Let $I_0 = [a, b]$. Now I_0 can be represented as the union of two intervals of equal length, viz. $I_0 = J_0 \cup K_0$, where $J_0 = [a, (a+b)/2]$ and $K_0 = [(a+b)/2, b]$. At least one of these intervals has an infinite intersection with X; denote this interval by I_1. If both $J_0 \cap X$ and $K_0 \cap X$ are infinite, then let $I_1 = J_0$. Repeating this procedure, represent I_1 as the union of two intervals J_1, K_1 of equal length, and let I_2 be one of these intervals having an infinite intersection with X; if both J_1 and K_1 have this property, choose J_1. We can repeat this construction *ad infinitum* to obtain a sequence (I_n) of closed intervals such that $I_{n+1} \subseteq I_n$ for all n. The length of I_n is $(b-a)/2^n$, which approaches 0 as $n \to \infty$ since $1/2^n \to 0$ as $n \to \infty$ (Example 7.11). Theorem 7.34 thus implies that there is a $c \in \mathbb{R}$ such that $c \in I_n$ for all n.

We now show that c is a limit point of X. Choose any $\delta > 0$. Since $(b-a)/2^n \to 0$ as $n \to \infty$ there is an N such that

$$\frac{b-a}{2^n} < \delta$$

for all $n > N$. Now each I_n by construction has an infinite intersection with X. For each $n > N$, the set I_n lies within the interval $(c-\delta, c+\delta)$ and hence $(c-\delta, c+\delta) \cap X$ is infinite. Consequently, c must be a limit point of X. □

Two results concerning the Bolzano-Weierstrass theorem should be noted.

(1) Theorem 7.35 establishes the existence of a limit point of X, but the limit point need not be an element of X.
(2) The set X may have an infinite number of limit points.

The set X of Example 7.3 illustrates the first remark; the set (a,b) of Example 7.4 illustrates the second remark.

Any set that contains an interior point *a fortiori* contains a limit point. The Bolzano-Weierstrass theorem is thus of particular interest when the set consists of isolated points. The range of a sequence produces such a set, and it is for sequences that the theorem is of conspicuous importance.

Corollary 7.36 *Any bounded sequence on $\mathbb{R}$ has a convergent subsequence.*

Proof. Let (a_n) be a bounded sequence on $\mathbb{R}$ and let

$$A = \{a_n : n \in \mathbb{N}\}.$$

If A is a finite set then there is a number $x \in A$ and an infinite subset $\{k_1, k_2, \dots\}$ of $\mathbb{N}$ with the properties that $a_{k_n} = x$ for all $n \in \mathbb{N}$ and $k_i < k_j$ whenever $i < j$. Then (a_{k_n}) is a constant subsequence of (a_n) and therefore converges.

Suppose that A is infinite. Since A is bounded, Theorem 7.35 shows that A must have a limit point x. The set $A \cap (x - 1/n, x + 1/n)$ is thus infinite for each $n \in \mathbb{N}$ and therefore there is a sequence (k_n) on $\mathbb{N}$ such that $k_n < k_{n+1}$ and $a_{k_n} \in (x - 1/n, x + 1/n)$ for all $n \in \mathbb{N}$. The proof of Lemma 7.24 can be used to show that $(a_{k_n}) \to x$ as $n \to \infty$. □

An alternative proof of Corollary 7.36 that avoids the nested intervals theorem rests on the fact that every sequence has a monotonic subsequence. This result is of interest in its own right and we prove it below.

Let (a_n) be a sequence on $\mathbb{R}$. A number $m \in \mathbb{N}$ is called a **peak index** for (a_n) if $a_n \leq a_m$ for all $n \geq m$. For example, consider the sequence $(1 + (-1)^n/n)$. The number 1 cannot be a peak index since $a_1 = 0$ and $a_2 = 3/2$. Since $a_n \leq 1 + 1/n \leq 1 + 1/2$ for all $n \geq 2$, we see that 2 is a peak index. The number 3 cannot be a peak index because $a_3 < a_4$, but 4 is a peak index since $a_n \leq 1 + 1/4$ for all $n \geq 4$.

Theorem 7.37 *Every sequence on $\mathbb{R}$ has a monotonic subsequence.*

Proof. Let (a_n) be a sequence on $\mathbb{R}$ and let $P \subseteq \mathbb{N}$ be the set of peak indices for (a_n). Then P is either finite or infinite.

Case I: If P is finite then there is an $N \in \mathbb{N}$ such that $m \leq N$ for all $m \in P$. In this case we can define a non-decreasing subsequence of (a_n). Let $k_1 = N + 1$. Now k_1 cannot be a peak index and therefore there is a $k_2 > k_1$ such that $a_{k_2} > a_{k_1}$. Thus for some $n \in \mathbb{N}$ we may assume the existence of numbers $k_1, k_2, \ldots, k_n$ such that

$$k_1 < k_2 < \cdots < k_n$$

and

$$a_{k_1} < a_{k_2} < \cdots < a_{k_n}.$$

Since $k_n > N$, k_n is not a peak index and hence there is a $k_{n+1} > k_n$ such that $a_{k_{n+1}} > a_{k_n}$. Then (a_{k_n}) is a non-decreasing subsequence of (a_n).

Case II: Suppose that P is infinite. For each $n \in \mathbb{N}$ let k_n be the nth peak index. Then $a_{k_{n+1}} \leq a_{k_n}$ for all $n \in \mathbb{N}$ and therefore the subsequence (a_{k_n}) is non-increasing. □

Theorem 7.37 provides an alternative proof of Corollary 7.36. Specifically, if (a_n) is bounded, then it is bounded above and bounded below. Theorem 7.37 implies that (a_n) has a monotonic subsequence (a_{k_n}). If (a_{k_n}) is non-decreasing, then Theorem 7.31 implies that (a_{k_n}) is convergent since it is bounded above; if (a_{k_n}) is non-increasing, then Theorem 7.31 implies that (a_{k_n}) is convergent since it is bounded below.

7.8 The Contraction Mapping Principle

In this section we prove another result that can be used to test a sequence for convergence. The result is a variant of a result called the contraction mapping principle, and it is particularly useful when a sequence is defined by a recurrence relation. It can also be used to prove the existence of solutions to certain equations.

Lemma 7.38 (**Geometric Series**) *Let (a_n) be the sequence on $\mathbb{R}$ defined by*

$$a_n = \sum_{k=0}^{n} q^k$$

for all $n \in \mathbb{N} \cup \{0\}$, where $q \in \mathbb{R}$ and $0 < q < 1$. Then (a_n) is a non-decreasing sequence and

$$\lim_{n \to \infty} a_n = \frac{1}{1-q}.$$

Proof. Evidently, $a_{n+1} - a_n = q^{n+1} > 0$ for all n and therefore (a_n) is non-decreasing. By Corollary 4.30 we have

$$a_n = \frac{1 - q^{n+1}}{1 - q}$$

for each non-negative integer n. Now $q^{n+1} \to 0$ as $n \to \infty$ (Example 7.8) and therefore

$$\lim_{n \to \infty} a_n = \frac{1}{1-q}.$$
□

Theorem 7.39 (**Contraction Mapping Principle**) *Let (x_n) be a sequence on $\mathbb{R}$, and suppose that there is a $q \in \mathbb{R}$ such that $0 < q < 1$ and*

$$|x_{n+1} - x_n| \leq q|x_n - x_{n-1}| \qquad (7.13)$$

for all $n \geq 1$. Then (x_n) converges.

Proof. We show that (x_n) is a Cauchy sequence and hence convergent by Theorem 7.14. First, we prove by induction that

$$|x_{n+1} - x_n| \leq q^n |x_1 - x_0| \qquad (7.14)$$

for all $n \geq 1$. If $n = 1$, then inequality 7.14 follows from inequality 7.13. Suppose that

$$|x_n - x_{n-1}| \leq q^{n-1}|x_1 - x_0| \tag{7.15}$$

for some integer $n > 1$. Then inequalities 7.13 and 7.15 imply that

$$|x_{n+1} - x_n| \leq q|x_n - x_{n-1}|$$
$$\leq q^n|x_1 - x_0|.$$

We thus conclude by induction that inequality 7.14 is satisfied for all $n \geq 1$. Hence we may assume that $x_1 \neq x_0$, for otherwise inequality 7.14 shows that $x_{n+1} = x_n$ for all $n \in \mathbb{N}$, so that (x_n) is a constant sequence and thus convergent.

For every $n \geq 1$ and every $k \geq 1$, the telescoping property shows that

$$|x_{n+k} - x_n| = \left|\sum_{j=1}^{k}(x_{n+j} - x_{n+j-1})\right|$$
$$\leq \sum_{j=1}^{k}|x_{n+j} - x_{n+j-1}|,$$

and using inequality 7.14 we have

$$|x_{n+k} - x_n| \leq \sum_{j=1}^{k} q^{n+j-1}|x_1 - x_0|$$
$$= |x_1 - x_0|q^n \sum_{j=1}^{k} q^{j-1}$$
$$= |x_1 - x_0|q^n \sum_{j=0}^{k-1} q^j.$$

Now $0 < q < 1$, and Lemma 7.38 implies that

$$\sum_{j=0}^{k-1} q^j \leq \frac{1}{1-q}.$$

Therefore,

$$|x_{n+k} - x_n| \leq \frac{q^n}{1-q}|x_1 - x_0|.$$

Recall that $q^n \to 0$ as $n \to \infty$. Choose $\epsilon > 0$. There exists $N \in \mathbb{N}$ such that
$$q^n < \frac{\epsilon(1-q)}{|x_1 - x_0|}$$
for all $n > N$. Choose $m, n \in \mathbb{N}$ such that $m > n > N$. Taking $k = m - n \geq 1$, we find that
$$|x_m - x_n| = |x_{n+k} - x_n| \leq \frac{q^n}{1-q}|x_1 - x_0| < \epsilon.$$
We conclude that (x_n) is a Cauchy sequence. Hence, by Theorem 7.14, (x_n) is convergent. $\square$

Example 7.14 Let (x_n) be the sequence defined by
$$x_0 = 1,$$
and
$$x_n = \frac{1 + x_{n-1}}{2 + x_{n-1}}$$
for all $n \geq 1$. We considered this sequence in Example 7.10, where it was shown that (x_n) is a non-increasing sequence that is bounded below by zero and hence convergent by Theorem 7.31. We now use Theorem 7.39 to prove convergence without appealing to monotonicity. For all $n \geq 1$ we have
$$|x_{n+1} - x_n| = \frac{|x_n - x_{n-1}|}{(2 + x_n)(2 + x_{n-1})}$$
$$\leq \frac{1}{4}|x_n - x_{n-1}|.$$
The sequence (x_n) thus converges by Theorem 7.39 with $q = 1/4$.

We can glean more information from this sequence. Since (x_n) is convergent and $2 + x_n > 0$ for all n, Theorem 7.20 implies that
$$\lim_{n \to \infty} x_n = \lim_{n \to \infty} \left(\frac{1 + x_{n-1}}{2 + x_{n-1}}\right)$$
$$= \frac{1 + \lim_{n \to \infty} x_{n-1}}{2 + \lim_{n \to \infty} x_{n-1}}.$$
Let $r = \lim_{n \to \infty} x_n$. Then $x_{n-1} \to r$ as $n \to \infty$; hence, r satisfies the equation
$$r = \frac{1 + r}{2 + r},$$

and so $r^2 + r - 1 = 0$. We can thus use this sequence to prove that there is a solution $r \in \mathbb{R}$ to the equation

$$x^2 + x - 1 = 0.$$

△

Let A be a function that maps elements of a set $X \subseteq \mathbb{R}$ into X. A point $x \in X$ is called a **fixed point** of A if $x = A(x)$.

If a convergent sequence (x_n) on X is defined in terms of an "initial value" x_0 and a recurrence relation of the form

$$x_n = A(x_{n-1})$$

for all $n \geq 1$, then the limit r of the sequence is a fixed point of A provided $r \in X$ and

$$\lim_{n \to \infty} A(x_n) = A(\lim_{n \to \infty} x_n).$$

Under these circumstances the sequence (x_n) can thus be used to prove that the equation

$$A(x) - x = 0$$

has a solution. Note that if the set X is closed then $r \in X$, since r is either a member of the sequence (x_n) or a limit point of X.

The construction of the real numbers was motivated in part by the observation that the equation $x^2 = 2$ does not have a solution in the rational numbers. Corollary 7.2 shows that this equation has a solution in the real numbers. If $w \in \mathbb{R}$, it is natural to enquire whether the equation $x^2 = w$ has a solution. Evidently, $x^2 \geq 0$, so that there is no solution in the real numbers if $w < 0$. This observation motivates the construction of the complex numbers in the next chapter. If $w \geq 0$, however, then we can show that $x^2 = w$ has a real solution. Before we state this result formally, a few remarks to motivate the proof are in order.

We will prove that $x^2 = w$ has a solution for $w \geq 0$ by showing that a suitable function A has a fixed point. The choice of A is to some extent arbitrary, but we require that any solution to $x = A(x)$ is also a solution to $x^2 = w$. The equation $x^2 = w$ is equivalent to

$$2x^2 = x^2 + w,$$

and for $x \neq 0$ this equation is equivalent to
$$x = \frac{1}{2}\left(x + \frac{w}{x}\right).$$
In this manner we are led to consider the function
$$A(x) = \frac{1}{2}\left(x + \frac{w}{x}\right)$$
for all $x \neq 0$. Any fixed point of A must be a solution to $x^2 = w$.

Theorem 7.40 *Let w be a non-negative real number. There exists a unique non-negative real number r such that $r^2 = w$.*

Proof. If $w = 0$, then $r = 0$ solves the equation $r^2 = 0$. There are no nonzero solutions to this equation and therefore the solution is unique.

Suppose that $w > 0$, and let (x_n) be the sequence defined by
$$x_0 = 1$$
and
$$x_n = \frac{1}{2}\left(x_{n-1} + \frac{w}{x_{n-1}}\right)$$
for all $n \geq 1$. Evidently $x_n > 0$ if $x_{n-1} > 0$, because $w > 0$. Since $x_0 > 0$, we thus have $x_n > 0$ for all $n \geq 1$.

For all $n \geq 1$ we find that
$$|x_{n+1} - x_n| = \frac{1}{2}\left|x_n - x_{n-1} + w\left(\frac{1}{x_n} - \frac{1}{x_{n-1}}\right)\right|$$
$$= \frac{1}{2}\left|x_n - x_{n-1} - w\frac{x_n - x_{n-1}}{x_n x_{n-1}}\right|$$
$$= \frac{1}{2}|x_n - x_{n-1}|\left|1 - \frac{w}{x_n x_{n-1}}\right|.$$

We shall show that
$$\left|1 - \frac{w}{x_n x_{n-1}}\right| \leq 1.$$

Since
$$x_n x_{n-1} = \frac{1}{2}(x_{n-1}^2 + w), \tag{7.16}$$

we have

$$1 - \frac{w}{x_n x_{n-1}} = 1 - \frac{2w}{x_{n-1}^2 + w}$$

$$= \frac{x_{n-1}^2 - w}{x_{n-1}^2 + w}.$$

As $w > 0$ it follows that

$$|x_{n-1}^2 - w| \le |x_{n-1}^2| + |w| = x_{n-1}^2 + w,$$

and so

$$\left|1 - \frac{w}{x_n x_{n-1}}\right| = \frac{|x_{n-1}^2 - w|}{x_{n-1}^2 + w} \le 1.$$

Therefore

$$|x_{n+1} - x_n| \le \frac{1}{2}|x_n - x_{n-1}|.$$

Theorem 7.39 implies that (x_n) is convergent. Let $r = \lim_{n \to \infty} x_n$. Since $x_n > 0$ for all n we have $r \ge 0$ by Theorem 7.19.

Equation 7.16 and Theorem 7.20 imply that

$$2r^2 = \lim_{n \to \infty} 2x_n x_{n-1}$$
$$= \lim_{n \to \infty} (x_{n-1}^2 + w)$$
$$= r^2 + w.$$

The limit of (x_n) is therefore a non-negative real number that satisfies the equation $r^2 = w$. Since $w > 0$, we have $r > 0$.

Suppose that s is a non-negative real number such that $s^2 = w$ for $w > 0$. Then $s > 0$ and $s^2 = r^2 = w$, so that

$$0 = s^2 - r^2 = (s - r)(s + r). \tag{7.17}$$

Since $r > 0$, $s + r > 0$ and the above equation shows that $s - r = 0$, so that $s = r$. We thus conclude that r is the unique positive number such that $r^2 = w$. □

If $r^2 = w$, then $(-r)^2 = (-1)^2 r^2 = r^2 = w$ so that the equation $x^2 = w$ has two distinct solutions if $w > 0$. The uniqueness component of Theorem 7.40 precludes the existence of other solutions positive or negative, because if $r^2 = s^2$ then $s = \pm r$ by equation 7.17. For any $w \geq 0$, the unique non-negative number r such that $r^2 = w$ is called the **square root** of w and denoted by $\sqrt{w}$.

Exercises 7

(1) Let $q \in \mathbb{R}_Q$ and $r \in \mathbb{R}_I$. Prove that:
   (a) $r + q \in \mathbb{R}_I$;
   (b) $rq \in \mathbb{R}_I$, provided $q \neq 0$.
(2) Verify that the axioms for a commutative ring with unity are satisfied for $(\mathbb{R}, +, \cdot, 0, 1)$.
(3) Let p and q be any real numbers such that $p < q$. Show that there exists an irrational number r such that $p < r < q$.
(4) State and prove an analogue of Theorem 5.14 (the Archimedean property) for real numbers.
(5) Let S_1 and S_2 be sets in $\mathbb{R}$.
   (a) If S_1 and S_2 are open, prove that $S_1 \cup S_2$ and $S_1 \cap S_2$ are open.
   (b) If S_1 and S_2 are closed, prove that $S_1 \cup S_2$ and $S_1 \cap S_2$ are closed.
(6) Prove that a set $s \subseteq \mathbb{R}$ is closed if and only if its complement $\mathbb{R} - S$ is open.
(7) Let (a_n) be a monotonic sequence and define the sequence (b_n) by
$$b_n = \frac{a_1 + a_2 + \cdots + a_n}{n}$$
for all $n \in \mathbb{N}$. Show that (b_n) is monotonic.
(8) Show that if $x_n > 0$ and $x_{n+1} \leq K x_n$ for all $n \in \mathbb{N}$, where $0 < K < 1$, then $x_n \to 0$ as $n \to \infty$.
(9) Let (x_n) be the sequence defined by $x_0 = 1$, and
$$x_n = \sqrt{3 x_{n-1}}$$
for all $n \geq 1$.
   (a) Prove that (x_n) is convergent by showing that it is non-decreasing and bounded above by 3.
   (b) Find $\lim_{n \to \infty} x_n$.
(10) Let $n \in \mathbb{N}$ and $x \in \mathbb{R}$, where $x > -1$. Prove that
$$(1 + x)^n \geq 1 + nx.$$
This inequality is known as **Bernoulli's inequality**.

(11) Let (b_n) be the sequence defined in Example 7.13.
   (a) Prove that (b_n) is a non-decreasing sequence.
   (b) Prove that the sequence (c_n) defined by
   $$c_n = \left(1 + \frac{1}{n}\right) b_n$$
   for $n \geq 1$ is a non-increasing sequence.
   (c) Use parts (a) and (b) to show that (b_n) is a convergent sequence.

(12) Prove that the sequence $(\sum_{k=1}^{n} 1/k)$ is divergent by showing that if it converges to a number s then both the sequences $(\sum_{k=1}^{n} 1/(2k))$ and $(\sum_{k=1}^{n} 1/(2k-1))$ converge to $s/2$.

(13) Let
$$s_n = \sum_{k=1}^{n} \frac{1}{k^2}$$
for all $n \in \mathbb{N}$. Show that
$$1 \leq s_n < 2 - \frac{1}{n}$$
for all $n \in \mathbb{N}$, and deduce that (s_n) is convergent.

(14) Let $A : X \to X$ be a function that maps the closed set $X \subseteq \mathbb{R}$ into X. The function A is called a **contraction** if there exists q such that $0 < q < 1$ and
$$|A(x) - A(y)| \leq q|x - y|$$
for all $x, y \in X$. Use Theorem 7.39 to prove that if A is a contraction, then there exists a unique fixed point for A in X. (This is a version of the contraction mapping principle.)

Chapter 8

Complex Numbers

8.1 Definition

For any real number x we have $x^2 \geq 0$. Thus the equation $x^2 = -1$, for example, has no real solution for x. We are therefore led to the development of yet another kind of number.

A **complex number** is an ordered pair of real numbers. The set of all complex numbers is denoted by $\mathbb{C}$. Thus $\mathbb{C} = \mathbb{R} \times \mathbb{R}$. Moreover we write $0 = (0,0)$ and $1 = (1,0)$. Thus the symbols "0" and "1" are pressed into service once more. The new uses for them will be reconciled with the old later. Note that $0 \neq 1$. We also define $i = (0,1)$.

8.2 Addition and Multiplication

Let $u, v, x, y \in \mathbb{R}$. We define
$$(u,v) + (x,y) = (u+x, v+y)$$
and
$$(u,v) \cdot (x,y) = (ux - vy, uy + vx).$$
We usually write $(u,v)(x,y)$ instead of $(u,v) \cdot (x,y)$. Note that
$$i^2 = (0,1)(0,1) = (-1, 0).$$

With the definitions above, we can show that the complex numbers form a field.

Theorem 8.1 $(\mathbb{C}, +, \cdot, 0, 1)$ *is a field.*

Proof. Choose $((t,u),(v,w),(x,y)) \in \mathbb{C} \times \mathbb{C} \times \mathbb{C}$. We have

$$(v,w) + (x,y) = (v+x, w+y)$$
$$= (x+v, y+w)$$
$$= (x,y) + (v,w).$$

Thus commutativity of addition for complex numbers follows immediately from that for real numbers. The associativity of addition for complex numbers follows similarly. Moreover

$$(x,y) + (0,0) = (x,y),$$

and

$$(x,y) + (-x,-y) = (x-x, y-y)$$
$$= (0,0).$$

Next,

$$(v,w)(x,y) = (vx - wy, vy + wx)$$
$$= (xv - yw, xw + yv)$$
$$= (x,y)(v,w),$$

$$(t,u)((v,w)(x,y)) = (t,u)(vx - wy, vy + wx)$$
$$= (t(vx-wy) - u(vy+wx), t(vy+wx) + u(vx-wy))$$
$$= (tvx - twy - uvy - uwx, tvy + twx + uvx - uwy)$$
$$= ((tv-uw)x - (tw+uv)y, (tw+uv)x + (tv-uw)y)$$
$$= (tv-uw, tw+uv)(x,y)$$
$$= ((t,u)(v,w))(x,y)$$

and

$$(t,u)((v,w) + (x,y)) = (t,u)(v+x, w+y)$$
$$= (t(v+x) - u(w+y), t(w+y) + u(v+x))$$
$$= (tv + tx - uw - uy, tw + ty + uv + ux)$$
$$= (tv - uw, tw + uv) + (tx - uy, ty + ux)$$
$$= (t,u)(v,w) + (t,u)(x,y).$$

We have now established that $(\mathbb{C}, +, \cdot, 0)$ is a commutative ring. Since

$$(x,y)(1,0) = (x - 0, 0 + y)$$
$$= (x,y),$$

$(\mathbb{C}, +, \cdot, 0, 1)$ is a commutative ring with unity.

Finally, suppose that $(x,y) \neq 0 = (0,0)$. Then either $x \neq 0$ or $y \neq 0$. Thus $x^2 > 0$ or $y^2 > 0$, and so $x^2 + y^2 > 0$ in any case. Hence

$$\left(\frac{x}{x^2+y^2}, -\frac{y}{x^2+y^2}\right) \in \mathbb{R} \times \mathbb{R} = \mathbb{C}.$$

Moreover

$$(x,y)\left(\frac{x}{x^2+y^2}, -\frac{y}{x^2+y^2}\right) = \left(\frac{x^2}{x^2+y^2} + \frac{y^2}{x^2+y^2}, -\frac{xy}{x^2+y^2} + \frac{xy}{x^2+y^2}\right)$$
$$= (1,0)$$
$$= 1,$$

and the proof is complete. □

Observe that if $(v,x) \in \mathbb{R} \times \mathbb{R}$ then

$$(v,0) + (x,0) = (v+x, 0)$$

and

$$(v,0)(x,0) = (vx - 0, 0 + 0)$$
$$= (vx, 0).$$

Let us define $\Phi(x) = (x,0)$ for each $x \in \mathbb{R}$. The calculations above show that $\Phi(v) + \Phi(x) = \Phi(v+x)$ and $\Phi(v)\Phi(x) = \Phi(vx)$. It is also clear that Φ is an injection from $\mathbb{R}$ into $\mathbb{C}$. This function thus sets up a $1:1$ correspondence between $\mathbb{R}$ and a subset of $\mathbb{C}$. We may therefore identify the real numbers with the complex numbers whose second component is 0. Moreover $\Phi(0) = (0,0) = 0$ and $\Phi(1) = (1,0) = 1$. Hence the two meanings for "0" and for "1" have been reconciled. Note also that $i^2 = (-1,0) = -1$.
If $(x,y) \in \mathbb{R} \times \mathbb{R}$, we see that

$$x + iy = (x,0) + (0,1)(y,0)$$
$$= (x,0) + (0,y)$$
$$= (x,y).$$

In future we shall usually write $x+iy$ instead of (x,y). We call x and y the **real** and **imaginary** parts, respectively, of $x+iy$, and write $x = \text{Re}\,(x+iy)$ and $y = \text{Im}\,(x+iy)$. If u, v, x, y are real, then

$$u + iv + x + iy = (u,v) + (x,y)$$
$$= (u+x, v+y)$$
$$= u + x + i(v+y).$$

and
$$\begin{aligned}(u+iv)(x+iy) &= (u,v)(x,y) \\ &= (ux-vy, uy+vx) \\ &= ux-vy+i(uy+vx) \\ &= ux+i^2vy+iuy+ivx.\end{aligned}$$

We infer that the addition and multiplication of complex numbers, written in the form $x+iy$ where x and y are real, can be effected by the algebraic laws for the manipulation of real numbers together with the rule that
$$i^2 = -1.$$
Moreover for any complex numbers z_1 and z_2 we have
$$\operatorname{Re}(z_1+z_2) = \operatorname{Re}(z_1) + \operatorname{Re}(z_2)$$
and
$$\operatorname{Im}(z_1+z_2) = \operatorname{Im}(z_1) + \operatorname{Im}(z_2).$$
Note also that because $(x,y) + (-x,-y) = (0,0)$ we have
$$\begin{aligned}-(x+iy) &= -(x,y) \\ &= (-x,-y) \\ &= -x - iy.\end{aligned}$$
Similarly if $(x,y) \neq 0$ then
$$\begin{aligned}\frac{1}{x+iy} &= \frac{1}{(x,y)} \\ &= \left(\frac{x}{x^2+y^2}, -\frac{y}{x^2+y^2}\right) \\ &= \frac{x}{x^2+y^2} - \frac{iy}{x^2+y^2} \\ &= \frac{x-iy}{x^2+y^2}.\end{aligned}$$
This result is also deduced from the computation
$$\begin{aligned}(x+iy)(x-iy) &= x^2 - (iy)^2 \\ &= x^2 - i^2 y^2 \\ &= x^2 + y^2\end{aligned}$$
by dividing both sides of the resulting equation by $(x+iy)(x^2+y^2)$.

8.3 Conjugates

Let $z = x + iy$ where x and y are real. We define $\bar{z} = x - iy$. Thus $z + \bar{z} = 2x = 2\text{Re}(z)$ and $z\bar{z} = x^2 + y^2$. We call $\bar{z}$ the **conjugate** of z. Note that $\bar{\bar{z}} = z$.

Since a complex number is merely an ordered pair of real numbers, it may be represented by a point in the xy-plane. Its conjugate is then represented by the point obtained upon reflection about the x-axis. Geometrical observations such as this play no rôle in the theory, and therefore do not require rigorous justification, but they may aid the reader's intuition.

Let $w = u + iv$ and $z = x + iy$, where u, v, x, y are real. The elementary properties of conjugates are now established by simple calculations. First,

$$\overline{w + z} = \overline{u + x + i(v + y)}$$
$$= u + x - i(v + y)$$
$$= u - iv + x - iy$$
$$= \bar{w} + \bar{z}$$

and

$$\overline{-z} = \overline{-x - iy}$$
$$= -x + iy$$
$$= -(x - iy)$$
$$= -\bar{z}.$$

Thus

$$\overline{w - z} = \overline{w + (-z)}$$
$$= \bar{w} + \overline{-z}$$
$$= \bar{w} + (-\bar{z})$$
$$= \bar{w} - \bar{z}.$$

Moreover

$$\overline{wz} = \overline{(u + iv)(x + iy)}$$
$$= \overline{ux - vy + i(uy + vx)}$$
$$= ux - vy - i(uy + vx)$$
$$= (u - iv)(x - iy)$$
$$= \bar{w}\bar{z}.$$

Finally if $z \neq 0$ then $w = wz/z$, and so the previous result yields $\bar{w} = \overline{w/z} \cdot \bar{z}$. Since $z \neq 0$ we have $\bar{z} \neq 0$, and division by $\bar{z}$ therefore gives $\overline{w/z} = \bar{w}/\bar{z}$.

8.4 Modulus

Since $i^2 = -1 < 0$, the field $(\mathbb{C}, +, \cdot, 0, 1)$ cannot be made into an ordered integral domain. (Recall that for each element z in an ordered integral domain we must have $z^2 \geq 0$.) Nevertheless it is possible to generalize the notion of absolute value. The resulting concept is called the modulus of a complex number.

Let $z = x + iy$ where x and y are real. The **modulus**, $|z|$, of z is defined as $\sqrt{x^2 + y^2}$. For example, $|i| = 1$. If z is real, then $y = 0$, so that $|z| = \sqrt{x^2} = |x|$, in agreement with the notation used earlier. Note that $x^2 + y^2 \geq 0$ and hence Theorem 7.40 ensures that the modulus is well-defined for all $z \in \mathbb{C}$. Observe that $|z| \geq 0$, and that $|z| = 0$ if and only if $z = 0$. Moreover $|z|^2 = x^2 + y^2 = z\bar{z}$ and $|\bar{z}| = |-z| = |z|$.

If $z \neq 0$, then $|z|$ may be interpreted geometrically in the xy-plane as the distance between the point representing z and the origin.

We conclude this section with some elementary properties of the modulus of a complex number.

Theorem 8.2 *For any $z \in \mathbb{C}$ the following inequalities are satisfied:*

(a) $|\operatorname{Re}(z)| \leq |z|$;
(b) $|\operatorname{Im}(z)| \leq |z|$.

Proof. Let $z = x + iy$, where x and y are real numbers. Then

$$|\operatorname{Re}(z)|^2 = x^2 \leq x^2 + y^2 = |z|^2.$$

Since $|\operatorname{Re}(z)|$ and $|z|$ are non-negative, the first inequality follows upon taking square roots. The second inequality can be established in a similar way. □

Theorem 8.3 *Let $(w, z) \in \mathbb{C} \times \mathbb{C}$. Then $|wz| = |w||z|$.*

Proof. We have

$$|wz|^2 = wz \cdot \overline{wz} = wz\bar{w}\bar{z} = w\bar{w}z\bar{z} = |w|^2|z|^2 = (|w||z|)^2.$$

Since both $|wz|$ and $|w||z|$ are non-negative, the required result follows by taking square roots. □

Corollary 8.4 *If $z \neq 0$ then*

$$\left|\frac{w}{z}\right| = \frac{|w|}{|z|}.$$

Proof. Since
$$|w| = \left|\frac{wz}{z}\right| = \left|\frac{w}{z}\right||z|,$$
the result follows upon division by $|z|$. □

The next theorem is the triangle inequality.

Theorem 8.5 *Let $(w, z) \in \mathbb{C} \times \mathbb{C}$. Then $|w + z| \leq |w| + |z|$.*

Proof. We have
$$\begin{aligned} |w + z|^2 &= (w + z)\overline{w + z} \\ &= (w + z)(\bar{w} + \bar{z}) \\ &= w\bar{w} + w\bar{z} + z\bar{w} + z\bar{z} \\ &= |w|^2 + w\bar{z} + \overline{w\bar{z}} + |z|^2 \\ &= |w|^2 + 2\text{Re}\,(w\bar{z}) + |z|^2 \\ &\leq |w|^2 + 2|\text{Re}\,(w\bar{z})| + |z|^2 \\ &\leq |w|^2 + 2|w\bar{z}| + |z|^2 \\ &= |w|^2 + 2|w||\bar{z}| + |z|^2 \\ &= |w|^2 + 2|w||z| + |z|^2 \\ &= (|w| + |z|)^2. \end{aligned}$$

Since $|w + z| \geq 0$ and $|w| + |z| \geq 0$, the result follows by taking square roots. □

Corollary 8.6 *Let $(w, z) \in \mathbb{C} \times \mathbb{C}$. Then $|w - z| \geq ||w| - |z||$.*

Proof. It suffices to show that $|w - z| \geq |w| - |z|$ and $|w - z| \geq |z| - |w|$. But the second inequality follows from the first since $|w - z| = |z - w|$. The first inequality is immediate from the fact that
$$\begin{aligned} |w| &= |w - z + z| \\ &\leq |w - z| + |z| \end{aligned}$$
by the triangle inequality. □

Corollary 8.7 *Let $z \in \mathbb{C}$. Then $|z| \leq |\text{Re}\,(z)| + |\text{Im}\,(z)|$.*

Proof. Let $z = x + iy$, where x and y are real. Then the triangle inequality implies that $|z| \leq |x| + |iy| = |x| + |y| = |\text{Re}\,(z)| + |\text{Im}\,(z)|$. □

8.5 Complex Sequences

Much of the mathematical machinery developed for sequences on $\mathbb{Q}$ and $\mathbb{R}$ can be readily adapted to sequences on $\mathbb{C}$. The definition of convergence, for instance, is formally similar to that for sequences on $\mathbb{R}$. Specifically, a sequence (z_n) on $\mathbb{C}$ is **convergent** if there is a $z \in \mathbb{C}$ with the property that for any positive real number ϵ there is an $N \in \mathbb{N}$ such that $|z_n - z| < \epsilon$ whenever $n > N$. The number z is called the **limit** of the sequence (z_n) and the relationship is denoted by $\lim_{n\to\infty} z_n = z$ or simply $z_n \to z$ as $n \to \infty$. We also say that the sequence **converges** to z.

The set
$$D(z;\epsilon) = \{w \in \mathbb{C} : |w - z| < \epsilon\}$$
can be interpreted geometrically as the set of points in $\mathbb{C}$ that are less than ϵ in distance from the point z. The set $D(z;\epsilon)$ can thus be regarded as a disc of radius ϵ centred at z. The points that are exactly a distance ϵ from z are excluded by the strict inequality so that the disc does not include the circle that forms its boundary. If $z_n \to z$ as $n \to \infty$, then the definition of convergence implies that, given a disc of any radius $\epsilon > 0$ centred at z, all but a finite number of terms in the sequence must lie within that disc.

Example 8.1 Use the definition of convergence to prove that the sequence (i^n/n) converges to 0.

Solution: We need to show that for any choice of $\epsilon > 0$ there is an N such that $|i^n/n - 0| = |i^n/n| < \epsilon$ whenever $n > N$. Choose $\epsilon > 0$. Now $|i^n/n| = |i^n|/|n| = 1/n$, since $|i| = 1$. Choose any $N > 1/\epsilon$. Then for any $n > N$ we have that $|i^n/n| = 1/n < 1/N < \epsilon$ and thus the sequence converges to 0.

△

Example 8.2 Use the definition of convergence to show that the sequence
$$\left(\frac{n - i^n}{2n^2 + i^n}\right)$$
converges to 0.

Solution: Following the strategy of the previous example note that
$$\left|\frac{n - i^n}{2n^2 + i^n} - 0\right| = \frac{|n - i^n|}{|2n^2 + i^n|}.$$

Now, the triangle inequality implies that

$$|n - i^n| \le |n| + |i^n| = n + 1,$$

and Corollary 8.6 shows that

$$|2n^2 + i^n| \ge ||2n^2| - |i^n|| = 2n^2 - 1.$$

Therefore

$$\left|\frac{n - i^n}{2n^2 + i^n}\right| \le \frac{n+1}{2n^2 - 1}.$$

For any $n \in \mathbb{N}$ we have $n + 1 \le 2n$ and since $n^2 \ge 1$ for any $n \in \mathbb{N}$ it follows that $2n^2 - 1 \ge n^2$ for any $n \in \mathbb{N}$. We thus have the inequality

$$\left|\frac{n - i^n}{2n^2 + i^n}\right| \le \frac{2n}{n^2} = \frac{2}{n}.$$

Choose any $\epsilon > 0$ and let $N > 2/\epsilon$. Then for any $n > N$ we have

$$\left|\frac{n - i^n}{2n^2 + i^n}\right| \le \frac{2}{n} < 2\frac{\epsilon}{2} = \epsilon,$$

as required.

△

Let (z_n) be a sequence on $\mathbb{C}$. For each $n \in \mathbb{N}$ there exist real numbers $\operatorname{Re}(z_n)$ and $\operatorname{Im}(z_n)$ such that $z_n = \operatorname{Re}(z_n) + i\operatorname{Im}(z_n)$ and therefore the sequence (z_n) defines two real sequences $(\operatorname{Re}(z_n))$ and $(\operatorname{Im}(z_n))$. The next theorem is a fundamental result connecting the complex and corresponding real sequences.

Theorem 8.8 *Let (z_n) be a sequence on $\mathbb{C}$. Then the sequence (z_n) converges if and only if the sequences $(\operatorname{Re}(z_n))$ and $(\operatorname{Im}(z_n))$ converge. Moreover, $z_n \to z$ as $n \to \infty$ if and only if $\operatorname{Re}(z_n) \to \operatorname{Re}(z)$ and $\operatorname{Im}(z_n) \to \operatorname{Im}(z)$ as $n \to \infty$.*

Proof. Suppose first that $z_n \to z$ as $n \to \infty$. We first show that $\operatorname{Re}(z_n) \to \operatorname{Re}(z)$ as $n \to \infty$. Choose any $\epsilon > 0$. Since $z_n \to z$ as $n \to \infty$ there is an N such that $|z_n - z| < \epsilon$ for all $n > N$. For any $n > N$, Theorem 8.2 implies that

$$|\operatorname{Re}(z_n) - \operatorname{Re}(z)| = |\operatorname{Re}(z_n - z)| \le |z_n - z| < \epsilon.$$

Therefore $\text{Re}(z_n) \to \text{Re}(z)$ as $n \to \infty$ by the definition of convergence. A similar argument can be used to show that $\text{Im}(z_n) \to \text{Im}(z)$ as $n \to \infty$.

Suppose now that $\text{Re}(z_n) \to \text{Re}(z)$ and $\text{Im}(z_n) \to \text{Im}(z)$ as $n \to \infty$. We will show that $z_n \to z$ as $n \to \infty$. Choose any $\epsilon > 0$. Now the convergence of $(\text{Re}(z_n))$ implies that there is an $N_1 \in \mathbb{N}$ such that

$$|\text{Re}(z_n) - \text{Re}(z)| < \frac{\epsilon}{2} \tag{8.1}$$

for all $n > N_1$. Similarly, there is an $N_2 \in \mathbb{N}$ such that

$$|\text{Im}(z_n) - \text{Im}(z)| < \frac{\epsilon}{2} \tag{8.2}$$

for all $n > N_2$. Let $N = \max\{N_1, N_2\}$. Then inequalities 8.1 and 8.2 are both satisfied for any $n > N$. For any $n \in \mathbb{N}$, Corollary 8.7 implies that

$$|z_n - z| \leq |\text{Re}(z_n) - \text{Re}(z)| + |\text{Im}(z_n) - \text{Im}(z)|;$$

therefore, for all $n > N$ we have that

$$|z_n - z| < \frac{\epsilon}{2} + \frac{\epsilon}{2} = \epsilon.$$

Hence $z_n \to z$ as $n \to \infty$. $\square$

A sequence (z_n) on $\mathbb{C}$ is a **Cauchy** sequence if for any positive real number ϵ there is an $N \in \mathbb{N}$ such that

$$|z_n - z_m| < \epsilon$$

for all $m, n > N$.

Theorem 8.9 *A sequence (z_n) on $\mathbb{C}$ is a Cauchy sequence if and only if $(\text{Re}(z_n))$ and $(\text{Im}(z_n))$ are Cauchy sequences.*

Proof. The proof of this result is similar to that of Theorem 8.8. Suppose that (z_n) is a Cauchy sequence. We will show that $(\text{Re}(z_n))$ is a Cauchy sequence. Choose any $\epsilon > 0$. Since (z_n) is a Cauchy sequence there is an N such that $|z_n - z_m| < \epsilon$ for all $m, n > N$. Now, by Theorem 8.2 we have that

$$|\text{Re}(z_n) - \text{Re}(z_m)| = |\text{Re}(z_n - z_m)| \leq |z_n - z_m| < \epsilon$$

for all $m, n > N$ and therefore $(\text{Re}(z_n))$ is a Cauchy sequence. The proof that $(\text{Im}(z_n))$ is a Cauchy sequence is similar.

Suppose now that $(\text{Re}(z_n))$ and $(\text{Im}(z_n))$ are Cauchy sequences. We will show that (z_n) is a Cauchy sequence. Choose any $\epsilon > 0$. Since $(\text{Re}(z_n))$ and $(\text{Im}(z_n))$ are Cauchy sequences there exists an $N \in \mathbb{N}$ such that

$$|\text{Re}(z_n) - \text{Re}(z_m)| < \frac{\epsilon}{2}$$

and

$$|\text{Im}(z_n) - \text{Im}(z_m)| < \frac{\epsilon}{2}$$

for all $m, n > N$. Corollary 8.7 implies that

$$|z_n - z_m| \leq |\text{Re}(z_n) - \text{Re}(z_m)| + |\text{Im}(z_n) - \text{Im}(z_m)|$$

for all $m, n \in \mathbb{N}$ and therefore

$$|z_n - z_m| < \frac{\epsilon}{2} + \frac{\epsilon}{2} = \epsilon$$

for all $m, n > N$. Therefore (z_n) is a Cauchy sequence by definition. □

Although a substantial number of results are needed to prove that the real number system is complete, once this is established the completeness of the complex number system follows readily.

Theorem 8.10 *A sequence on $\mathbb{C}$ is convergent if and only if it is a Cauchy sequence.*

Proof. Theorems 8.9, 7.14 and 8.8 provide a string of equivalences that prove the above theorem. Specifically, Theorem 8.9 shows that (z_n) is a Cauchy sequence if and only if $(\text{Re}(z_n))$ and $(\text{Im}(z_n))$ are Cauchy sequences. Theorem 7.14 implies that $(\text{Re}(z_n))$ and $(\text{Im}(z_n))$ are Cauchy sequences if and only if $(\text{Re}(z_n))$ and $(\text{Im}(z_n))$ both converge. Finally, Theorem 8.8 shows that $(\text{Re}(z_n))$ and $(\text{Im}(z_n))$ both converge if and only if (z_n) converges. □

The key to the above proof is Theorem 7.14, which required significant mathematical effort to establish. The other results used in the above proof are comparatively simple to prove. Roughly speaking, the complex number system "inherits" its completeness property from the real numbers and this makes the proof of completeness simple.

Results such as the uniqueness of a limit for a convergent sequence can be readily extended to sequences on $\mathbb{C}$ using Theorem 8.8.

Theorem 8.11 *If a sequence on $\mathbb{C}$ converges, then the limit of the sequence is unique.*

Proof. Suppose that the sequence (z_n) on $\mathbb{C}$ converges to z. Theorem 8.8 shows that the sequences $(\text{Re}(z_n))$ and $(\text{Im}(z_n))$ are convergent and that $\text{Re}(z_n) \to \text{Re}(z)$ and $\text{Im}(z_n) \to \text{Im}(z)$ as $n \to \infty$. Theorem 7.16 shows that these real sequences have unique limits and hence (z_n) must have a unique limit. □

Theorem 8.8 can be used to establish extensions of Theorems 7.17, 7.20 and 7.22 to sequences on $\mathbb{C}$. The proofs are straightforward applications of Theorem 8.8 and we omit them.

Theorem 8.12 *Let (z_n) be a sequence on $\mathbb{C}$ and suppose that $z_n \to z$ as $n \to \infty$. Then every subsequence (z_{k_n}) of (z_n) is convergent and $z_{k_n} \to z$ as $n \to \infty$.*

Theorem 8.13 *If (z_n) and (w_n) are sequences on $\mathbb{C}$ and $z_n \to z$, $w_n \to w$ as $n \to \infty$, then*

(a) $\lim_{n \to \infty}(z_n + w_n) = z + w$;
(b) $\lim_{n \to \infty} z_n w_n = zw$;
(c) $\lim_{n \to \infty} z_n/w_n = z/w$, *provided* $w \neq 0$.

Theorem 8.14 *If (z_n) is a sequence on $\mathbb{C}$ and $z_n \to z$ as $n \to \infty$, then $|z_n| \to |z|$ as $n \to \infty$.*

A sequence (z_n) on $\mathbb{C}$ is **bounded** if there is an $M \in \mathbb{R}$ such that $|z_n| < M$ for all $n \in \mathbb{N}$.

Theorem 8.15 *Any convergent sequence on $\mathbb{C}$ is bounded.*

Proof. Suppose that (z_n) is convergent. Then Theorem 8.8 implies that that the sequences $(\text{Re}(z_n))$ and $(\text{Im}(z_n))$ are convergent. Theorem 7.18 shows that these real sequences are therefore bounded. Hence there are numbers $M_1, M_2 \in \mathbb{R}$ such that $|\text{Re}(z_n)| < M_1$ and $|\text{Im}(z_n)| < M_2$ for all $n \in \mathbb{N}$. Corollary 8.7 therefore implies that

$$|z_n| \leq |\text{Re}(z_n)| + |\text{Im}(z_n)| < M_1 + M_2$$

for all $n \in \mathbb{N}$ and hence (z_n) is bounded. □

Note that Theorem 7.19 does not have a close analogue for complex sequences because $(\mathbb{C}, +, \cdot, 0, 1)$ is not an ordered field.

8.6 Sets in $\mathbb{C}$

The notions of open sets, limit points and closed sets can be extended readily to sets in the complex plane. The analogue of the open interval $(x - r, x + r)$ in $\mathbb{R}$ of length $2r > 0$ centred at x is the **open disc**

$$D(w; r) = \{z \in \mathbb{C} : |z - w| < r\}$$

of radius r (diameter $2r$) centred at w; the analogue of the closed interval $[x - r, x + r]$ in $\mathbb{R}$ is the **closed disc**

$$\bar{D}(w; r) = \{z \in \mathbb{C} : |z - w| \leq r\}.$$

We also define the set

$$\partial D(w; r) = \{z \in \mathbb{C} : |z - w| = r\}.$$

Geometrically, $\partial D(w; r)$ is a circle of radius r centred at w.

Let X be a subset of $\mathbb{C}$. A point $z \in X$ is called an **interior point** of X if there is a $\delta > 0$ such that $D(z; \delta) \subseteq X$. A set $X \subseteq \mathbb{C}$ is called **open** if every element of X is an interior point of X.

Example 8.3 Prove that for $w \in \mathbb{C}$ and $r > 0$, the set $D(w; r)$ is open.

Solution: Let $z \in D(w; r)$ and $d = |z - w|$. Then $d < r$ and Theorem 7.12 implies that there is a $\delta \in \mathbb{R}$ such that $0 < \delta < r - d$. For any $u \in D(z; \delta)$,

$$\begin{aligned} |u - w| &= |u - z + z - w| \\ &\leq |u - z| + |z - w| \\ &\leq \delta + d \\ &< r; \end{aligned}$$

hence, $D(z; \delta) \subseteq D(w; r)$ and thus z is an interior point of $D(w; r)$. Since z is an arbitrary point of $D(w; r)$, we conclude that every point of $D(w; r)$ is an interior point. Therefore $D(w; r)$ is an open set.

$\triangle$

Example 8.4 Prove that the set

$$\Pi_0 = \{z \in \mathbb{C} : \operatorname{Re}(z) > 0\}$$

is open.

Solution: Let $z \in \Pi_0$. Then $\text{Re}(z) > 0$, so that by Theorem 7.12 there is a $\delta > 0$ such that $0 < \delta < \text{Re}(z)$. Let $u \in D(z; \delta)$. Then

$$\delta > |u - z| \geq \text{Re}(z - u) = \text{Re}(z) - \text{Re}(u),$$

and so we have $\text{Re}(u) > \text{Re}(z) - \delta > 0$. Therefore, $u \in \Pi_0$ and consequently $D(z; \delta) \subseteq \Pi_0$. Hence, z is an interior point of Π_0. Since $z \in \Pi_0$ is arbitrary, we conclude that every point in Π_0 is an interior point and hence that Π_0 is an open set. It is called the **right open half plane.**

$\triangle$

An element z of a set $X \subseteq \mathbb{C}$ is said to be **isolated** if there is a $\delta > 0$ such that

$$D(z; \delta) \cap X = \{z\}.$$

A number z is called a **limit point** of a set $X \subseteq \mathbb{C}$ if for every $\delta > 0$ we have

$$D(z; \delta) \cap (X - \{z\}) \neq \emptyset.$$

The comments concerning limit points for sets in $\mathbb{R}$ apply to sets in $\mathbb{C}$. Specifically, z need not be a member of X to be a limit point of X, any interior point of X is a limit point of X, and no isolated point of X can be a limit point of X.

A set $X \subseteq \mathbb{C}$ is called **closed** if X contains all its limit points. The **closure** $\bar{X}$ of a set X is the union of X with the set of its limit points. As in the case of a set of real numbers, the closure of a set of complex numbers is a closed set.

Example 8.5 Prove that if $w \in \mathbb{C}$ and $r > 0$ then the set $\bar{D}(w; r)$ is closed.

Solution: Let z be a limit point of $\bar{D}(w; r)$ and suppose that $z \notin \bar{D}(w; r)$. Then $|z - w| > r$ and hence by Theorem 7.12 there is a $\delta > 0$ such that $|z - w| - r > \delta > 0$. Let $u \in D(z; \delta)$. Then $|u - z| < \delta$ and thus

$$\begin{aligned} |u - w| &= |u - z - (w - z)| \\ &\geq ||u - z| - |w - z|| \\ &\geq |z - w| - |u - z| \\ &> |z - w| - \delta \\ &> r, \end{aligned}$$

so that $u \notin \bar{D}(w;r)$. Therefore $D(z;\delta) \cap \bar{D}(w;r) = \emptyset$ and hence z cannot be a limit point of $\bar{D}(w;r)$. We thus conclude that if z is a limit point of $\bar{D}(w;r)$, then $z \in \bar{D}(w;r)$. The set $\bar{D}(w;r)$ is therefore closed.

△

Example 8.6 Find the closure of the set

$$A = \left\{ \frac{i^n}{n} : n \in \mathbb{N} \right\}.$$

Solution: The reader can verify that every point of A is isolated (cf. Example 7.3). If A has any limit points, they are not members of A. Example 8.1 shows that $i^n/n \to 0$ as $n \to \infty$. Thus for any $\delta > 0$ there is an N such that $|i^n/n| < \delta$ for all $n > N$. Hence, for any $\delta > 0$ we have $D(0;\delta) \cap A \neq \emptyset$, and since $0 \notin A$ it follows that 0 is a limit point of A. Arguing as in Example 7.3, we see that there are no other limit points of A. The closure of A is therefore $A \cup \{0\}$.

△

A set $X \subseteq \mathbb{C}$ is **bounded** if there is an $M \in \mathbb{R}$ such that $|z| < M$ for all $z \in X$. If X is closed and bounded then it is called **compact**. For example, the set $\bar{D}(w;r)$ is compact for any $w \in \mathbb{C}$ and $r > 0$.

The Bolzano-Weierstrass Theorem can be extended to complex sets. Rather than pursue an extension of Theorem 7.35, we shall prove the complex extension of Corollary 7.36.

Theorem 8.16 (**Bolzano – Weierstrass**) *Every bounded sequence on $\mathbb{C}$ has a convergent subsequence.*

Proof. Let (z_n) be a bounded sequence on $\mathbb{C}$. There is an M such that $|z_n| < M$ for all $n \in \mathbb{N}$, and therefore, by Theorem 8.2, $|\text{Re}(z_n)| < M$ and $|\text{Im}(z_n)| < M$ for all $n \in \mathbb{N}$. The sequences $(\text{Re}(z_n))$ and $(\text{Im}(z_n))$ are thus bounded. Corollary 7.36 implies that there is a convergent subsequence $(\text{Re}(z_{k_n}))$ of $(\text{Re}(z_n))$; therefore, there is a subsequence (z_{k_n}) of (z_n) such that $(\text{Re}(z_{k_n}))$ converges. Let $w_n = z_{k_n}$ for each $n \in \mathbb{N}$. The sequence $(\text{Im}(w_n))$ is bounded and therefore by Corollary 7.36 there is a convergent subsequence $(\text{Im}(w_{l_n}))$ of $(\text{Im}(w_n))$. Hence, there is a subsequence (w_{l_n}) of (w_n) such that $(\text{Im}(w_{l_n}))$ converges. Since $(\text{Re}(w_n))$ converges, Theorem 7.17 implies that $(\text{Re}(w_{l_n}))$ converges; consequently, by Theorem 8.8 the sequence (w_{l_n}) converges. □

8.7 Quadratic Equations

Theorem 7.40 shows that every non-negative real number has a square root. On the other hand, the equation $x^2 = w$ does not have a solution in the real numbers if $w < 0$, but it is easy to show that this equation has a solution in the complex numbers.

Lemma 8.17 *Let $w \in \mathbb{R}$. Then there is an $r \in \mathbb{C}$ such that $r^2 = w$.*

Proof. If $w \geq 0$ then the result follows by Theorem 7.40. Suppose that $w < 0$. Then $|w| > 0$, and $w = -|w|$. Theorem 7.40 implies that there is a $\rho \in \mathbb{R}$ such that $\rho^2 = |w|$. Let $r = i\rho$. Then $r^2 = -\rho^2 = -|w| = w$. □

It is natural to enquire whether the equation $z^2 = w$ is soluble for z in $\mathbb{C}$ whenever $w \in \mathbb{C}$. We show that this equation is indeed soluble in the complex numbers. First, however, we make a few remarks to motivate the proof and establish a simple lemma.

Let $w = x + iy$, where $x, y \in \mathbb{R}$, and suppose that there is an $r \in \mathbb{C}$ such that $r^2 = x + iy$. Lemma 8.17 guarantees the existence of such an r if $y = 0$. Suppose that $y \neq 0$. Let $r = a + ib$, where $a, b \in \mathbb{R}$. Since $r^2 = w$ and $\text{Im}(w) = y \neq 0$, we have $b \neq 0$, so that r can be expressed in the form

$$r = b(\tau + i),$$

where $\tau = a/b$. Hence,

$$r^2 = b^2(\tau + i)^2 = x + iy,$$

and therefore

$$\text{Re}(r^2) = b^2(\tau^2 - 1) = x$$

and

$$\text{Im}(r^2) = b^2 2\tau = y.$$

The last two equations and the condition $y \neq 0$ imply that $\tau y = 2b^2 \tau^2 > 0$ and

$$\frac{x}{y} = \frac{\tau^2 - 1}{2\tau}.$$

The above arguments suggest that we can construct a number $r \in \mathbb{C}$ such that $r^2 = w$, for if $\text{Im}(w) \neq 0$ then there is a $\tau \in \mathbb{R}$ such that $\tau y > 0$

and
$$\tau^2 - 1 = 2\frac{\tau x}{y}.$$
In this case we could choose $b \in \mathbb{R}$ such that
$$b^2 = \frac{y}{2\tau},$$
and hence $\text{Im}(r^2) = b^2 2\tau = y$ and $\text{Re}(r^2) = b^2(\tau^2 - 1) = b^2 2\tau x/y = x$.

Lemma 8.18 *For any $\gamma \in \mathbb{R}$ there are numbers $\tau_1, \tau_2 \in \mathbb{R}$ such that $\tau_1 > 0$, $\tau_2 < 0$ and*
$$\tau_k^2 + 2\gamma\tau_k - 1 = 0$$
for $k = 1, 2$.

Proof. For any $\gamma \in \mathbb{R}$ and any $\tau \in \mathbb{R}$
$$\tau^2 + 2\gamma\tau - 1 = \tau^2 + 2\gamma\tau + \gamma^2 - \gamma^2 - 1$$
$$= (\tau + \gamma)^2 - (\gamma^2 + 1).$$

Since $\gamma^2 + 1 > 0$, Theorem 7.40 implies that there is a positive $r \in \mathbb{R}$ such that $r^2 = \gamma^2 + 1$. Let $\tau_1 = r - \gamma$. Then
$$\tau_1^2 + 2\gamma\tau_1 - 1 = r^2 - 2r\gamma + \gamma^2 + 2\gamma r - 2\gamma^2 - 1$$
$$= r^2 - \gamma^2 - 1$$
$$= 0$$
and
$$\tau_1 = \sqrt{\gamma^2 + 1} - \gamma > 0.$$
Since $r^2 = \gamma^2 + 1$, we have $(-r)^2 = \gamma^2 + 1$. Let $\tau_2 = -r - \gamma$. Then
$$\tau_2^2 + 2\gamma\tau_2 - 1 = 0$$
and
$$\tau_2 = -\sqrt{\gamma^2 + 1} - \gamma < 0.$$
$\square$

Theorem 8.19 *For any $w \in \mathbb{C}$ there is an $r \in \mathbb{C}$ such that $r^2 = w$.*

Proof. Let $w = x + iy$, where $x, y \in \mathbb{R}$. If $y = 0$, then the result follows from Lemma 8.17. Suppose that $y \neq 0$. Lemma 8.18 implies that the equation

$$\tau^2 - 2\frac{x\tau}{y} - 1 = 0$$

has a positive solution and a negative solution among the real numbers. Let τ_0 be the solution to this equation such that $\tau_0 y > 0$. Since $y/(2\tau_0) > 0$, Theorem 7.40 implies that there is a $b > 0$ such that

$$b^2 = \frac{y}{2\tau_0}.$$

Let $r = b(\tau_0 + i)$. Then

$$r^2 = b^2(\tau_0^2 - 1) + ib^2 2\tau_0.$$

The definition of τ_0 shows that

$$\tau_0^2 - 1 = 2\tau_0 \frac{x}{y}.$$

Hence

$$\operatorname{Re}(r^2) = b^2(\tau_0^2 - 1) = \frac{y}{2\tau_0} 2\tau_0 \frac{x}{y} = x,$$

and

$$\operatorname{Im}(r^2) = b^2 2\tau_0 = y,$$

so that $r^2 = x + iy = w$. □

The proofs of Lemma 8.18 and Theorem 8.19 can also be used to express the solution r in terms of x and y. We leave it to the reader to verify the solution

$$r = \frac{x + \sqrt{x^2 + y^2} + iy}{\sqrt{2\left(x + \sqrt{x^2 + y^2}\right)}}.$$

Consider now the general quadratic equation

$$a_2 z^2 + a_1 z + a_0 = 0,$$

where $a_k \in \mathbb{C}$ for $k = 0, 1, 2$ and $a_2 \neq 0$. Since $a_2 \neq 0$, this equation is equivalent to

$$z^2 + 2c_1 z + c_0 = 0,$$

where $2c_1 = a_1/a_2$ and $c_0 = a_0/a_2$. Theorem 8.19 can be used to show that this equation is soluble in the complex numbers for any $c_1, c_0 \in \mathbb{C}$. Specifically, we have the identity

$$z^2 + 2c_1 z + c_0 = z^2 + 2c_1 z + c_1^2 - c_1^2 + c_0$$
$$= (z + c_1)^2 - c_1^2 + c_0.$$

Theorem 8.19 implies that there is an $s \in \mathbb{C}$ such that $s^2 = c_1^2 - c_0$. Let $r = s - c_1$. Then

$$(r + c_1)^2 = s^2 = c_1^2 - c_0,$$

and hence

$$r^2 + 2c_1 r + c_0 = 0.$$

The general quadratic equation is thus soluble in $\mathbb{C}$ for any choice of coefficients $a_k \in \mathbb{C}$, provided that $a_2 \neq 0$. If $a_2 = 0$ and $a_1 \neq 0$ then we have the solution $z = -a_0/a_1$.

Having established that the general quadratic equation is soluble in $\mathbb{C}$ it is natural to ask whether the general cubic equation

$$a_3 z^3 + a_2 z^2 + a_1 z + a_0 = 0$$

is soluble in $\mathbb{C}$ for any choice of coefficients $a_k \in \mathbb{C}$ such that $a_0 = 0$ if $a_k = 0$ for each $k > 0$. More generally, we ask whether the equation

$$a_n z^n + a_{n-1} z^{n-1} + \cdots + a_1 z + a_0 = 0,$$

where $n \geq 1$, has solutions in $\mathbb{C}$ for any choice of $a_k \in \mathbb{C}$ satisfying the condition $a_n \neq 0$. It turns out that such equations are indeed soluble in $\mathbb{C}$. We prove this fact in the next chapter.

Exercises 8

(1) Prove that

$$\operatorname{Re}\left(\frac{w+z}{w-z}\right) = \frac{|w|^2 - |z|^2}{|w-z|^2}$$

for any complex numbers z, w such that $z \neq w$.
(2) Prove that $|\operatorname{Re}(z)| + |\operatorname{Im}(z)| \leq \sqrt{2}|z|$ for any $z \in \mathbb{C}$.
(3) Prove that

(a) $|1 - \bar{z}w|^2 - |z - w|^2 = (1 - |z|^2)(1 - |w|^2)$,
(b) $|z + iw|^2 + |w + iz|^2 = 2(|z|^2 + |w|^2)$

for any complex numbers z and w.

(4) Use the definition of convergence to prove that the sequence
$$\left(\frac{1 + i^n n}{n^2}\right)$$
converges.

(5) Let $N : \mathbb{C} \to \mathbb{R}$ be a function and suppose that there exist positive real numbers α and β such that
$$\alpha|z| \le N(z) \le \beta|z|$$
for all $z \in \mathbb{C}$. Prove that (z_n) is a Cauchy sequence if and only if for each $\epsilon > 0$ there is an n_0 such that $N(z_n - z_m) \le \epsilon$ whenever $n, m > n_0$. If $N(z) = |\text{Re}(z)| + |\text{Im}(z)|$ find such an α and β.

(6) Let S_1 and S_2 be open sets in $\mathbb{C}$. Prove that $S_1 \cup S_2$ and $S_1 \cap S_2$ are open sets.

(7) Suppose that (z_n) is a bounded divergent sequence. Prove that (z_n) has two subsequences that converge to different limits.

(8) If $z_n \to z$ as $n \to \infty$, show that $|z_n| \to |z|$ as $n \to \infty$.

(9) If $z_n \to z$ as $n \to \infty$, prove that the sequence (w_n), defined by
$$w_n = \frac{1}{n}(z_1 + z_2 + \cdots + z_n)$$
for all $n \in \mathbb{N}$, converges to z.

Chapter 9

The Fundamental Theorem of Algebra

In this chapter we show that every non-constant polynomial $P(z)$ over $\mathbb{C}$ takes the value 0 for some $z \in \mathbb{C}$. This result is called the fundamental theorem of algebra.

9.1 General Results Concerning Polynomials

For any $n \in \mathbb{N} \cup \{0\}$ a **polynomial** of **degree** n is a function $P_n : \mathbb{C} \to \mathbb{C}$ such that

$$P_n(z) = a_n z^n + a_{n-1} z^{n-1} + \cdots + a_1 z + a_0$$

for all $z \in \mathbb{C}$, where $a_k \in \mathbb{C}$ for $k = 0, 1, \ldots, n$ and $a_n \neq 0$. If $n = 0$, then we say that the polynomial is **constant**. For succinctness, we use the notation

$$P_n(z) = \sum_{k=0}^{n} a_k z^k,$$

with the convention that $0^0 = 1$. In this section we prove some properties of polynomials that will be needed in Section 9.4.

Theorem 9.1 (Continuity) *For any $w \in \mathbb{C}$ and any $\epsilon > 0$ there is a $\delta > 0$ such that*

$$|P_n(z) - P_n(w)| < \epsilon$$

for all $z \in D(w; \delta)$.

Proof. Choose $w \in \mathbb{C}$ and $\epsilon > 0$. If $n = 0$, then P_n is constant and the result is trivial. Suppose that $n \geq 1$. For any $k \geq 1$ and $z \in \mathbb{C}$ we have, by

Theorem 4.29,
$$z^k - w^k = (z-w)\sum_{j=1}^{k} z^{k-j}w^{j-1},$$
and hence
$$|z^k - w^k| \le |z-w|\sum_{j=1}^{k}|z|^{k-j}|w|^{j-1}$$
by the triangle inequality. Corollary 8.6 shows that for any $z \in D(w;1)$ we have
$$|z| - |w| \le ||z|-|w|| \le |z-w| < 1,$$
so that $|z| < 1 + |w|$. For each $k \in \{1, 2, \ldots, n\}$ let
$$\begin{aligned}\lambda_k(w) &= \sum_{j=1}^{k}(1+|w|)^{k-j}|w|^{j-1} \\ &= (1+|w|)^{k-1} + \sum_{j=2}^{k}(1+|w|)^{k-j}|w|^{j-1} \\ &> 0.\end{aligned}$$
Then for each $z \in D(w;1)$ and each $k \in \{1,2,\ldots,n\}$ we have
$$|z^k - w^k| \le |z-w|\lambda_k(w).$$
Now, since $z^0 - w^0 = 1 - 1 = 0$, for any such z we have
$$\begin{aligned}|P_n(z) - P_n(w)| &= \left|\sum_{k=1}^{n} a_k(z^k - w^k)\right| \\ &\le \sum_{k=1}^{n}|a_k||z^k - w^k| \\ &\le |z-w|\sum_{k=1}^{n}|a_k|\lambda_k(w) \\ &= |z-w|C(w),\end{aligned}$$
where C is the function from $\mathbb{C}$ into $\mathbb{R}$ given by
$$C(z) = \sum_{k=1}^{n}|a_k|\lambda_k(z)$$

for each $z \in \mathbb{C}$. Note that $C(w) > 0$ since $a_n \neq 0$ and $\lambda_k(w) > 0$ for each $k = 1, \ldots, n$. Let $\delta = \min\{1, \epsilon/C(w)\}$. Then, for all $z \in D(w; \delta)$,

$$|P_n(z) - P_n(w)| \leq |z - w|C(w) < \frac{\epsilon}{C(w)}C(w) = \epsilon.$$

□

Roughly speaking, Theorem 9.1 shows that if $|z - w|$ is "small", then $|P_n(z) - P_n(w)|$ is also "small". A more sophisticated interpretation of this statement in terms of open sets is given in Exercises 9 No. 1. The next two results follow readily from Theorem 9.1. The first can also be proved directly from Theorem 8.13.

Corollary 9.2 *Let (z_k) be a convergent sequence on $\mathbb{C}$ and let*

$$z = \lim_{k \to \infty} z_k.$$

Then

$$\lim_{k \to \infty} P_n(z_k) = P_n(z).$$

Proof. Choose $\epsilon > 0$. Let C be as defined in the proof of Theorem 9.1. Since $z_k \to z$ as $k \to \infty$, there is an N such that $|z_k - z| < \min\{1, \epsilon/C(z)\}$ for all $k > N$. Choose $k > N$. Then $|z_k - z| < 1$, and so

$$|P_n(z_k) - P_n(z)| \leq |z_k - z|C(z) < \frac{\epsilon}{C(z)}C(z) = \epsilon.$$

The sequence $(P_n(z_k))$ therefore converges to $P_n(z)$. □

Corollary 9.3 *Let (z_k) be a convergent sequence on $\mathbb{C}$ and let*

$$z = \lim_{k \to \infty} z_k.$$

Then

$$\lim_{k \to \infty} |P_n(z_k)| = |P_n(z)|.$$

Proof. Corollary 9.2 shows that $\lim_{k \to \infty} P_n(z_k) = P_n(z)$, and the result follows from Theorem 8.14. □

Theorem 9.1 shows essentially that P_n maps open sets to open sets (Exercises 9 No. 1). The next result shows that P_n maps compact sets in $\mathbb{C}$ to compact sets in $\mathbb{C}$.

Theorem 9.4 *Let $X \subseteq \mathbb{C}$ be a compact set and let*

$$\mathcal{P}_X = \{|P_n(z)| : z \in X\}.$$

Then $\mathcal{P}_X$ is a compact set.

Proof. The set X is compact and therefore bounded. Hence there is an $M > 0$ such that $|z| < M$ for all $z \in X$. Evidently,

$$P_n(z) \leq \sum_{k=0}^{n} |a_k||z|^k < \sum_{k=0}^{n} |a_k|M^k$$

for all $z \in X$ and therefore $\mathcal{P}_X$ is bounded.

It remains to show that $\mathcal{P}_X$ is closed. Suppose that x is a limit point of $\mathcal{P}_X$. Then by Lemma 7.24 there is a sequence (x_k) in $\mathcal{P}_X$ such that $x_k \to x$ as $k \to \infty$. Since $x_k \in \mathcal{P}_X$, there is a $z_k \in X$ such that $|P_n(z_k)| = x_k$. Thus there is a sequence (z_k) in X such that $x_k = |P_n(z_k)|$ for each k. The set X is bounded and hence the sequence (z_k) is bounded. The Bolzano-Weierstrass theorem (Theorem 8.16) thus implies that some subsequence (z_{k_j}) of (z_k) converges to a limit z. Since X is closed, $z \in X$, for if there is no $k \in \mathbb{N}$ such that $z_k = z$ then z must be a limit point of X. Corollaries 9.2 and 9.3 imply that

$$|P_n(z)| = \lim_{j \to \infty} |P_n(z_{k_j})| = \lim_{j \to \infty} x_{k_j} = x.$$

Now, $z \in X$ and so $x \in \mathcal{P}_X$. Hence $\mathcal{P}_X$ contains all its limit points and is therefore closed. □

The next result shows that the range of a non-constant polynomial is not a bounded set.

Theorem 9.5 *Let $n \geq 1$. For any $M \geq 0$, there is an $R > 0$ such that*

$$|P_n(z)| > M$$

for all $z \in \mathbb{C} - D(0; R)$.

Proof. Choose $M \geq 0$. For any $z \in \mathbb{C} - D(0;1)$ we have $|z| \geq 1$. Since $a_n \neq 0$, Corollary 8.6 gives

$$|P_n(z)| = |a_n z^n + \sum_{k=0}^{n-1} a_k z^k|$$

$$\geq |a_n z^n| - \left|\sum_{k=0}^{n-1} a_k z^k\right|$$

$$\geq |a_n||z|^n - \sum_{k=0}^{n-1} |a_k||z|^k$$

$$\geq |a_n||z|^n - |z|^{n-1} \sum_{k=0}^{n-1} |a_k|$$

$$= |a_n||z|^n \left(1 - \frac{1}{|a_n||z|} \sum_{k=0}^{n-1} |a_k|\right)$$

$$= |a_n||z|^n \left(1 - \frac{K}{|z|}\right),$$

where

$$K = \frac{1}{|a_n|} \sum_{k=0}^{n-1} |a_k|.$$

Choose z so that $|z| > \max\{1, 2K\}$. Then $|z|^n \geq |z|$ (since $n \geq 1$) and $K/|z| < 1/2$, so that

$$|P_n(z)| \geq |a_n||z|^n \left(1 - \frac{K}{|z|}\right)$$

$$> \frac{|a_n||z|}{2}.$$

Now let $R = \max\{1, 2K, 2M/|a_n|\}$. Then, for any $z \in \mathbb{C} - D(0;R)$,

$$|P_n(z)| > \frac{|a_n||z|}{2} \geq \frac{|a_n|}{2} \frac{2M}{|a_n|} = M.$$

□

9.2 Polynomials with Real Coefficients

In this section we focus on polynomials of the form

$$T_n(x) = \sum_{k=0}^{n} a_k x^k,$$

where $a_k \in \mathbb{R}$ for $k = 1, 2, \ldots, n$, and $a_n \neq 0$. Evidently, these polynomials are a special case of the general polynomial of degree n. They are called **polynomials with real coefficients**. The continuity property of polynomials (Theorem 9.1) is exploited to show that polynomials of odd degree with real coefficients take the value 0 for some real x. We establish this result at the end of the section after proving two lemmas. The first lemma shows that if T_n is nonzero at a point $c \in \mathbb{R}$, then it is nonzero in some open interval containing c.

Lemma 9.6 *If $T_n(c) > 0$ for some $c \in \mathbb{R}$, then there is a $\delta > 0$ such that $T_n(x) > 0$ for all $x \in (c - \delta, c + \delta)$.*

Proof. Suppose $T_n(c) > 0$. Theorem 9.1 implies that there is a $\delta > 0$ such that

$$|T_n(x) - T_n(c)| < \frac{1}{2} T_n(c)$$

for all $x \in (c - \delta, c + \delta)$. Therefore

$$-\frac{1}{2} T_n(c) < T_n(x) - T_n(c),$$

and hence

$$0 < \frac{1}{2} T_n(c) < T_n(x)$$

for all $x \in (c - \delta, c + \delta)$. □

Lemma 9.7 *Let n be an odd integer such that $n \geq 3$ and let $a_n > 0$. Then there exist numbers $x_1 < 0$, $x_2 > 0$ and $N > 0$ such that*

$$T_n(x) < Nx < 0$$

for all $x < x_1$, and

$$T_n(x) > Nx > 0$$

for all $x > x_2$.

Proof. Let $A = \max\{|a_j| : j = 0, 1, \ldots, n\}$. Thus $A > 0$. For all $x \in \mathbb{R}$ such that $|x| > 1$ we have

$$T_n(x) = a_n x^n + \sum_{k=0}^{n-1} a_k x^k$$

$$\leq a_n x^n + \left|\sum_{k=0}^{n-1} a_k x^k\right|$$

$$\leq a_n x^n + \sum_{k=0}^{n-1} |a_k||x|^k$$

$$\leq a_n x^n + A \sum_{k=0}^{n-1} |x|^k$$

$$\leq a_n x^n + nA|x|^{n-1},$$

and if $x > 1$ then

$$T_n(x) \geq a_n x^n - \left|\sum_{k=0}^{n-1} a_k x^k\right|$$

$$\geq a_n x^n - nAx^{n-1}$$

$$\geq a_n x^{n-1}\left(x - \frac{nA}{a_n}\right).$$

Now n is odd, so that $n - 1$ is even and $x^{n-1} \geq 0$ for all $x \in \mathbb{R}$. Choose $x_1 = \min\{-1, -2nA/a_n\} < 0$. Then

$$x_1 + \frac{nA}{a_n} \leq -\frac{nA}{a_n},$$

and for all $x < x_1$ we have $|x| > 1$ and

$$T_n(x) \leq a_n x^n + nAx^{n-1}$$

$$= x^{n-1} a_n \left(x + \frac{nA}{a_n}\right)$$

$$< x^{n-1} a_n \left(x_1 + \frac{nA}{a_n}\right)$$

$$\leq x^{n-1} a_n \left(-\frac{nA}{a_n}\right)$$

$$= -nAx^{n-1}.$$

Since $x < -1$ and $n-2$ is odd and positive it follows that $x^{n-2} < -1$; hence $-x^{n-1} < x$ and so

$$T_n(x) < Nx < 0,$$

where $N = nA > 0$. Similarly, choose $x_2 = \max\{1, 2nA/a_n\} > 0$. Then

$$x - \frac{nA}{a_n} > x_2 - \frac{nA}{a_n} \geq \frac{nA}{a_n}$$

for all $x > x_2$ and hence

$$T_n(x) \geq a_n x^{n-1}\left(x - \frac{nA}{a_n}\right) > nAx^{n-1} > Nx > 0$$

for all $x > x_2$. □

Theorem 9.8 *Let n be an odd integer such that $n \geq 3$. Then there is a $c \in \mathbb{R}$ such that $T_n(c) = 0$.*

Proof. Since $T_n(x) = 0$ if and only if $-T_n(x) = 0$, we can assume without loss of generality that $a_n > 0$.
Let

$$X = \{x \in \mathbb{R} : T_n(x) < 0\}.$$

Lemma 9.7 implies that there is an $x_2 > 0$ such that $T_n(x) > 0$ for all $x > x_2$. Therefore X is bounded above by x_2. Lemma 9.7 also implies that there is an x_1 such that $T_n(x) < 0$ for all $x < x_1$ and hence X is not empty. Theorem 7.27 shows that X must have a least upper bound, c. We show that $T_n(c) = 0$.

Suppose that $T_n(c) > 0$. Since $c = \text{lub } X$, for any $\gamma > 0$ there is an $x \in (c - \gamma, c)$ such that $T_n(c) < 0$. But Lemma 9.6 implies that there is a $\delta > 0$ such that $T_n(x) > 0$ for all $x \in (c - \delta, c + \delta)$. This contradiction shows that $T_n(c) \leq 0$.

Suppose that $T_n(c) < 0$. Then we can apply Lemma 9.6 to the function $-T_n$ and thereby deduce the existence of a $\delta > 0$ such that $T_n(c) < 0$ for all $x \in (c - \delta, c + \delta)$. In particular, there is an $x > c$ such that $T_n(c) < 0$, in contradiction to the fact that c is an upper bound of X. We thus conclude that $T_n(c) \geq 0$. Therefore $T_n(c) = 0$. □

9.3 nth Roots of a Complex Number

In Section 8.7 we showed that for any $w \in \mathbb{C}$ there is an $r \in \mathbb{C}$ such that $r^2 = w$ (Theorem 8.19). In this section we generalize this result to show that for any natural number n and any $w \in \mathbb{C}$ there is an $r \in \mathbb{C}$ such that $r^n = w$. The number r is called an **nth root** of w. We first prove this result for the case where n is odd. The proof follows a strategy similar to that used to prove Theorem 8.19.

Lemma 9.9 *Let n be a positive odd integer and let $w \in \mathbb{C}$. Then there is an $r \in \mathbb{C}$ such that $r^n = w$.*

Proof. If $n = 1$ then we can choose $r = w$. Suppose $n \geq 3$ and let $w = a + ib$, where $a, b \in \mathbb{R}$. If $b = 0$, then Theorem 9.8 implies that there is a (real) solution to the equation

$$x^n - w = 0,$$

so that we can choose r as this solution.

Suppose that $b \neq 0$. For any $\tau \in \mathbb{R}$ the binomial theorem (Theorem 5.11) gives

$$(\tau + i)^n = \sum_{k=0}^{n} \binom{n}{k} \tau^{n-k} i^k,$$

and hence

$$\operatorname{Re}\left((\tau + i)^n\right) = \sum_{k=0}^{(n-1)/2} (-1)^k \binom{n}{2k} \tau^{n-2k}$$

and

$$\operatorname{Im}\left((\tau + i)^n\right) = \sum_{k=0}^{(n-1)/2} (-1)^k \binom{n}{2k+1} \tau^{n-2k-1}.$$

Therefore the function T_n defined by

$$T_n(\tau) = \operatorname{Re}\left((\tau + i)^n\right) - \frac{a}{b} \operatorname{Im}\left((\tau + i)^n\right)$$

is a polynomial of degree n with real coefficients. Theorem 9.8 implies that there is a $\tau_0 \in \mathbb{R}$ such that

$$T_n(\tau_0) = 0.$$

Hence
$$\operatorname{Re}\left((\tau_0+i)^n\right) = \frac{a}{b}\operatorname{Im}\left((\tau_0+i)^n\right). \tag{9.1}$$

Suppose that $a = 0$. Let $n = 2j+1$. Theorem 9.8 implies that there is a $\sigma \in \mathbb{R}$ such that $\sigma^n = b$. If j is even, then $(i\sigma)^n = i\sigma^n = ib = w$ so that we may take $r = i\sigma$. If j is odd, then $(-i\sigma)^n = -i^n\sigma^n = ib = w$ and we may take $r = -i\sigma$.

Suppose that $a \neq 0$. Since $\tau_0 \in \mathbb{R}$ and $|(\tau_0+i)^n| = |\tau_0+i|^n \neq 0$ we see that $\operatorname{Re}\left((\tau_0+i)^n\right)$ and $\operatorname{Im}\left((\tau_0+i)^n\right)$ cannot both be zero. Since $a \neq 0$, we conclude from 9.1 that $\operatorname{Re}\left((\tau_0+i)^n\right) \neq 0$. Theorem 9.8 implies that there is a $c \in \mathbb{R}$ such that
$$c^n = \frac{a}{\operatorname{Re}\left((\tau_0+i)^n\right)}.$$

Let $r = c(\tau_0+i)$. Then, using equation 9.1,
$$\begin{aligned} r^n &= c^n(\tau_0+i)^n \\ &= c^n\operatorname{Re}\left((\tau_0+i)^n\right) + ic^n\operatorname{Im}\left((\tau_0+i)^n\right) \\ &= a + ia\frac{\operatorname{Im}\left((\tau_0+i)^n\right)}{\operatorname{Re}\left((\tau_0+i)^n\right)} \\ &= a + ib \\ &= w. \end{aligned}$$
$\square$

Theorem 9.10 (**Nth Root of a Complex Number**) *Let n be a positive integer and let $w \in \mathbb{C}$. Then there is an $r \in \mathbb{C}$ such that $r^n = w$.*

Proof. Any positive integer can be written in the form
$$n = 2^k m,$$
where m is a positive odd integer and k is a non-negative integer. If $k = 0$, then the result follows from Lemma 9.9. Suppose $k \neq 0$. We first consider the case where $m = 1$. Theorem 8.19 implies that there is a $\sigma_1 \in \mathbb{C}$ such that
$$\sigma_1^2 = w.$$
Now suppose that $\sigma_{k-1}^{2^{k-1}} = w$ for some integer $k > 1$ and some $\sigma_{k-1} \in \mathbb{C}$. Then by Theorem 8.19 there exists a $\sigma_k \in \mathbb{C}$ such that $\sigma_k^2 = \sigma_{k-1}$. Hence
$$\sigma_k^n = \sigma_k^{2^k} = \sigma_k^{2 \cdot 2^{k-1}} = (\sigma_k^2)^{2^{k-1}} = \sigma_{k-1}^{2^{k-1}} = w.$$

Now suppose that $m > 1$. By the previous case there exists a $\sigma_k \in \mathbb{C}$ such that $\sigma_k^{2^k} = w$, and Lemma 9.9 implies that there is an $r \in \mathbb{C}$ such that $r^m = \sigma_k$. Thus

$$r^n = r^{2^k m} = (r^m)^{2^k} = \sigma_k^{2^k} = w.$$

$\square$

9.4 The Fundamental Theorem of Algebra

In this section we prove the fundamental theorem of algebra. We first establish a technical result and then prove a theorem concerning the minimum modulus of a polynomial in a disc. This result is of interest in its own right. Having established this theorem we then prove the fundamental theorem.

Lemma 9.11 *Let G be a non-constant polynomial such that $G(0) = 1$. For each $\delta > 0$ let*

$$\rho_\delta = \{|G(z)| : z \in D(0; \delta)\}.$$

Then, for any $\delta > 0$, 1 cannot be a lower bound for ρ_δ.

Proof. Choose $\delta > 0$. Since G is non-constant and $G(0) = 1$, we may write

$$G(z) = 1 + \sum_{k=j}^{n} a_k z^k$$

where $1 \leq j \leq n$, $a_k \in \mathbb{C}$ for all k, $a_j \neq 0$ and $a_n \neq 0$. Theorem 9.10 implies that there is an $r \in \mathbb{C}$ such that $r^j = -1/a_j$. Let $H(z) = G(rz)$ for all $z \in \mathbb{C}$. Since $r \neq 0$ it suffices to find an $x \in (0, \delta/|r|)$ such that $|H(x)| < 1$, for then

$$|rx| < |r|\frac{\delta}{|r|} = \delta.$$

Now, for any $x \in \mathbb{R}$,

$$H(x) = 1 + a_j (rx)^j + \sum_{k=j+1}^{n} a_k (rx)^k$$

$$= 1 - x^j + \sum_{k=j+1}^{n} b_k x^k,$$

where $b_k = a_k r^k$ for $k = j+1, \ldots, n$. For any $x \in \mathbb{R}$ such that $0 < x < 1$ and all positive integers m we have $0 < x^{m+1} < x^m < 1$, and therefore

$$|H(x)| \le |1 - x^j| + \sum_{k=j+1}^{n} |b_k| x^k$$

$$\le 1 - x^j + x^{j+1} \sum_{k=j+1}^{n} |b_k|$$

$$= 1 + x^j(xB - 1),$$

where

$$B = \sum_{k=j+1}^{n} |b_k| \ge 0.$$

If $B = 0$ then $|H(x)| < 1$ for any $x \in (0, 1)$. Suppose that $B > 0$, and let $\gamma = \min\{1, \delta/|r|, 1/B\}$. Then for any $x \in (0, \gamma)$ we have $xB - 1 < 0$ and $x^j > 0$; hence

$$|H(x)| \le 1 + x^j(xB - 1) < 1.$$

Therefore 1 cannot be a lower bound for ρ_δ. □

The next result shows that in a closed disc a non-constant polynomial assumes its minimum on the boundary of the disc and not in the interior, unless the minimum is zero.

Theorem 9.12 (**Minimum Modulus**) *Let P_n be a non-constant polynomial and $\delta > 0$. Then the set*

$$\mathcal{P}_D = \{|P_n(z)| : z \in \bar{D}(0; \delta)\}$$

has a greatest lower bound μ in $\mathcal{P}_D$. Moreover, either

(1) $\mu = 0$; or
(2) $|P_n(z)| > \mu$ for all $z \in D(0; \delta)$ and $|P_n(z_0)| = \mu$ for some $z_0 \in \partial D(0; \delta)$.

Proof. For any $\delta > 0$ the set $\bar{D}(0; \delta)$ is compact (cf. Example 8.5) and hence Theorem 9.4 implies that $\mathcal{P}_D$ is compact. Therefore, by Theorem 7.30, $\mathcal{P}_D$ has a greatest lower bound $\mu \in \mathcal{P}_D$. Thus there is a $z_0 \in \bar{D}(0; \delta)$ such that $|P_n(z_0)| = \mu$.

If $\mu = 0$, then there is nothing to prove. Suppose that $\mu \neq 0$ and that $z_0 \in D(0;\delta)$. Let $\lambda = P_n(z_0)$. Then $|\lambda| = \mu > 0$. Let

$$G(z) = \frac{P_n(z+z_0)}{\lambda}$$

for all $z \in \mathbb{C}$. Then G is a non-constant polynomial and $G(0) = \lambda/\lambda = 1$. Now z_0 is an interior point of $D(0;\delta)$; consequently, there is a $\gamma > 0$ such that $z + z_0 \in D(0;\delta)$ for all $z \in D(0;\gamma)$. Since $\mu = $ glb $\mathcal{P}_D$, the greatest lower bound for

$$\rho_\gamma = \{|G(z)| : z \in D(0;\gamma)\}$$

must be $\mu/|\lambda| = 1$. But $1 \neq $ glb ρ_γ by Lemma 9.11. We thus conclude that $z_0 \in \partial D(0;\delta)$. $\square$

Theorem 9.13 **(Fundamental Theorem of Algebra)** *Let P_n be a non-constant polynomial. Then there exists a $z_0 \in \mathbb{C}$ such that $P_n(z_0) = 0$.*

Proof. Since P_n is non-constant, Theorem 9.5 shows that there is an $R > 0$ such that $|P_n(z)| > |a_0|$ for all $z \in \mathbb{C} - D(0;R)$. Theorem 9.12 implies that the set

$$\mathcal{P}_D = \{|P_n(z)| : z \in \bar{D}(0;R)\}$$

has a greatest lower bound and that there is a $z_0 \in \bar{D}(0;R)$ such that $|P_n(z_0)| = $ glb $\mathcal{P}_D$. Moreover either $|z_0| = R$ or glb $\mathcal{P}_D = 0$. If $|z_0| = R$, then $|P_n(z_0)| > |a_0|$. But $|P_n(0)| = |a_0|$, so that $|P_n(z_0)|$ is not a lower bound of $\mathcal{P}_D$. This contradiction shows that $|z_0| \neq R$ and hence that glb $\mathcal{P}_D = 0$. Thus $|P_n(z_0)| = 0$ and so $P_n(z_0) = 0$. $\square$

Exercises 9

(1) Let $X \subseteq \mathbb{C}$ be a nonempty open set. Show that the set

$$\mathcal{P}_X = \{P_n(z) : z \in X\}$$

is an open set.

(2) Prove Corollary 9.2 using Theorem 8.13.
(3) Let G be as defined in Lemma 9.11. Prove that for any $\delta > 0$, $|G(0)|$ cannot be an upper bound for ρ_δ.
(4) Prove that in a closed disc a non-constant polynomial achieves its maximum modulus on the boundary of the disc and not in the interior.

(5) Let P_n be a non-constant polynomial and suppose that there is a number $R > 0$ such that $P_n(z) = 1$ whenever $|z| = R$. Prove that there is a $z_0 \in D(0; R)$ such that $P_n(z_0) = 0$.

(6) **(a)** Let P and D be polynomials and suppose that D is not constant. Show that there exist polynomials Q and R such that the degree of R is less than that of D and

$$P(z) = D(z)Q(z) + R(z)$$

for all $z \in \mathbb{C}$.

(b) If P is a non-constant polynomial and $w \in \mathbb{C}$, prove that $P(w) = 0$ if and only if there is a polynomial Q such that

$$P(z) = (z - w)Q(z)$$

for all $z \in \mathbb{C}$.

(c) Deduce that if

$$P(z) = \sum_{k=0}^{n} a_k z^k$$

for all $z \in \mathbb{C}$, where $n > 0$, then there exist $w_1, w_2, \ldots, w_n \in \mathbb{C}$ such that

$$P(z) = a_n(z - w_1)(z - w_2) \cdots (z - w_n)$$

for all $z \in \mathbb{C}$.

Index

nth power, 83
nth root, 215

absolute value, 88
addition, 49, 60, 78, 108
   associativity, 50
   commutativity, 51
   of Cauchy sequences, 133
additive identity, 78
additive inverse, 79
Archimedean Property, 121
associative law, 92
associativity, 73, 76
   of composition of functions, 37
Axiom of Choice, 28

base, 104
Bernoulli's Inequality, 184
bijection, 34, 39
Binomial Theorem, 116
Bolzano-Weierstrass Theorem, 175, 200
bounded above, 160, 162
bounded below, 160

cancellation law, 51, 82
cardinality, 67
Cartesian product, 31, 69
Cauchy
   Criterion, 152
   sequence, 132, 148, 195
Cauchy Criterion, 166

cell, 32
closed disc, 198
commutative, 60
commutative law, 96
commutative ring with unity, 142
commutativity, 73, 76
compact set, 165
completeness, 151, 196
complex number, 185
   imaginary part, 188
   real part, 188
composition of functions, 38
conclusion, 5
congruence, 105
conjugate, 189
conjunction
   of propositions, 4
contraction, 184
contraction mapping principle, 184
contrapositive, 6
converse
   of a proposition, 5
cube, 83

de Morgan's Laws, 10, 16
   for sets, 28
decimal, 104
density, 119, 121, 149, 151
disjunction
   of cases, 7
   of propositions, 3
distributive law, 98

distributivity of multiplication over
  addition, 76
division, 112
Division Theorem, 101
domain, 33
double negation, 9
dummy variable, 16

empty set
  uniqueness of, 22
equation, 20
equinumerous, 65
equivalence
  class, 33
  dual, 9
  relation, 32, 135
equivalence relation, 71
exponentiation, 54, 82, 113

factor, 63, 78
factorial, 64
field, 112, 186
  ordered, 118
function, 33
  composition of, 35
  identity, 36
  inverse of, 35
  n-admissible, 69
  one-to-one correspondence, 34

geometric series, 177
greater than, 56, 85, 86, 118
greatest lower bound, 161

hypothesis, 5

identity, 13, 95
  additive, 81
  multiplicative, 81
image, 34
implication, 5
  vacuous, 5
inclusion, 19
  proper, 20, 31
induction, 43, 60, 90
inductive

hypothesis, 43
  set, 28, 41
inequality, 56
injection, 34, 35, 38, 44
integer, 72
  even, 102
  odd, 102
integers
  addition, 73
  sum, 73
integral domain, 81, 112
  ordered, 86, 118, 146
interval
  closed, 156
  length, 156
  of natural numbers, closed, 64
  open, 156
  semi-infinite, 156
  semi-open, 156
inverse, 35
  multiplicative, 112
irrational number, 141

laws
  absorption, 13
  associative, 15
  commutative, 12
  de Morgan's, 10
  distributive, 13
least element, 59
least upper bound, 160, 162
less than, 56, 85, 86, 118
limit
  of a sequence, 128
limit point, 175
lower bound, 131, 160

maximal element, 160
minimal element, 160
Minimum Modulus Theorem, 218
modulus, 190
monoid, 95
monotonic, 166
multiplication, 52, 63, 76, 78, 109
  associativity, 54
  commutativity, 53

distibutivity over, 55
distributivity over addition, 52
of Cauchy sequences, 133
multiplicative
  inverse, 112

natural number, 42
  even, 70
  odd, 70
necessity, 5
negative, 84, 86, 118, 145
nested intervals, 174

open disc, 198
operation
  associative, 15
  binary, 39, 48, 52, 54
  commutative, 12
  distributive, 12
  unary, 39, 45, 48
order, 144, 146
ordered
  n-tuple, 68
  pair, 29
  triple, 30

partial ordering, 32
partition, 32, 33
  finer, 39
peak index, 176
point
  fixed, 180
  interior, 157, 198
  isolated, 157, 199
  limit, 157, 199
polynomial, 207
  constant, 207
positive, 84, 86, 118, 145
power, 54
predecessor, 44
predicate, 15
premise, 5
principle
  of double negation, 9
product, 52, 63, 76, 78, 109
proof

  by contradiction, 6
proposition, 1
  conjunction of, 4
  contrapositive of, 6
  converse, 5
  disjunction of, 3
  equivalence of, 7
  negation of, 3
  undecidable, 7

quantifier
  existential, 15
  universal, 16
quotient, 102, 112

range, 33
rational numbers, 107
real number system, 146
Recursion Theorem, 45, 60
relation
  antisymmetric, 31
  circular, 39
  equivalence, 32, 71, 107
  inverse of, 35
  reflexive, 31
  symmetric, 31
  transitive, 31
remainder, 102
representative, 33
ring, 78, 98
  commutative, 81
  with unity, 81

Sandwich Theorem, 170
semigroup, 60, 92
  commutative, 96
sequence
  bounded, 131, 176, 197
  Cauchy, 132, 138, 148, 195
  convergent, 128, 132, 146, 176, 193
  definition, 125
  divergent, 128, 139
  equivalent, 135
  induced, 147
  limit, 128, 193
  non-decreasing, 166

   non-increasing, 166
   null, 143
   positive, 144
set
   bounded, 175, 200
   closed, 159, 199
   closure, 199
   closure of, 159
   compact, 165, 200, 210
   complement of, 24
   containment in, 2
   disjoint, 23
   element of, 2
   empty, 2
   equal, 20
   finite, 65, 156
   inductive, 28
   infinite, 65, 156, 175
   intersection, 23
   intersection of, 28
   null, 2
   open, 157, 198
   partition of, 32
   power, 22
   singleton, 23
   symmetric difference, 39
   transitive, 43
   union of, 25
sign, 85, 118
square, 83
square root, 88, 183
subsequence, 130, 176
subset
   definition of, 19
   proper, 20
subtraction, 80
successor function, 42
sufficiency, 5
sum, 49, 63, 78, 108
summand, 63
surjection, 34, 38, 43
symmetric difference, 39

Telescoping Property, 99
term, 63, 78
   of a sequence, 125

theorem, 6
Triangle Inequality, 90, 192
Trichotomy Property, 56
truth table, 3
   for $p \Leftrightarrow q$, 7
   for $p \Rightarrow q$, 5
   for $p \vee q$, 4
   for $p \wedge q$, 4

unit element, 81
upper bound, 131, 160

Venn diagram, 21

DATE DUE

```
SCI QA 299.8 .L58 2003

Little, Charles H. C. 1947-

The number systems of
 analysis
```